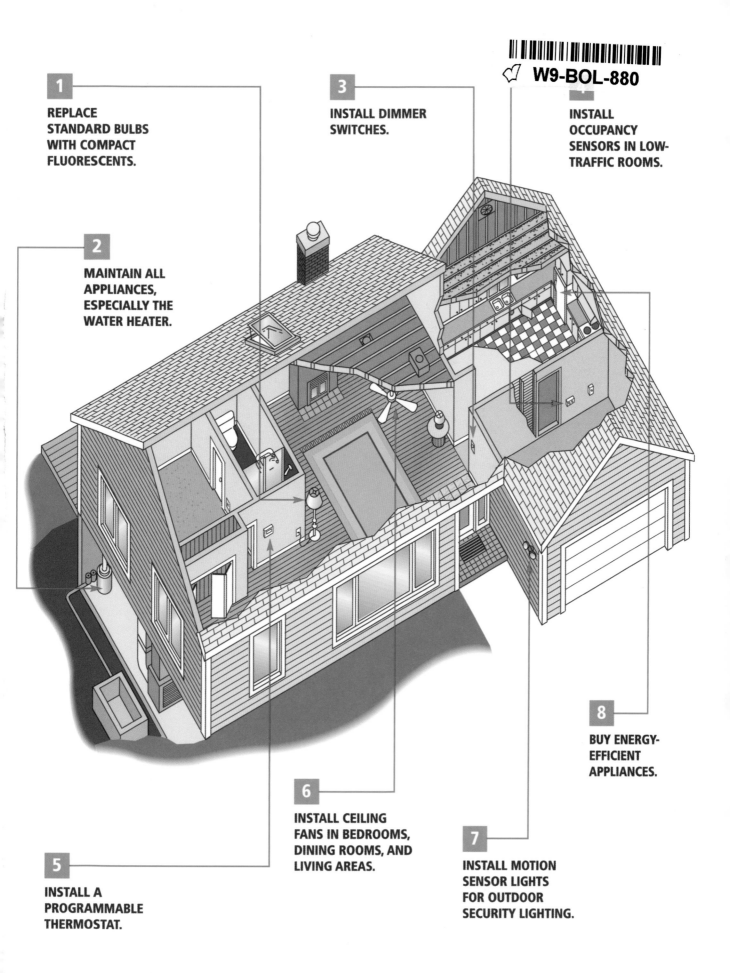

1 REPLACE STANDARD BULBS WITH COMPACT FLUORESCENTS.

2 MAINTAIN ALL APPLIANCES, ESPECIALLY THE WATER HEATER.

3 INSTALL DIMMER SWITCHES.

4 INSTALL OCCUPANCY SENSORS IN LOW-TRAFFIC ROOMS.

5 INSTALL A PROGRAMMABLE THERMOSTAT.

6 INSTALL CEILING FANS IN BEDROOMS, DINING ROOMS, AND LIVING AREAS.

7 INSTALL MOTION SENSOR LIGHTS FOR OUTDOOR SECURITY LIGHTING.

8 BUY ENERGY-EFFICIENT APPLIANCES.

Wiring 1-2-3.

Meredith® Books Development Team
Project Editors: Catherine M. Staub, John P. Holms
Art Director: John Eric Seid
Copy Chief: Catherine Hamrick
Copy and Production Editor: Terri Fredrickson
Managers, Book Production: Pam Kvitne, Marjorie J. Schenkelberg
Contributing Copy Editor: Margaret Smith
Contributing Proofreaders: John C. Edwards, Steve Hallam, Raymond L. Kast, Ralph Selzer
Contributing Designers: Tim Abramowitz, Joyce E. DeWitt
Electronic Production Coordinator: Paula Forest
Editorial Assistants: Renee E. McAtee, Karen Schirm

Greenleaf Publishing, Inc.
Editor: Dave Toht
Writer: Steve Cory
Art Director: Jean De Vaty
Associate Art Director: Rebecca Jon Michaels
Photography: Dan Stultz
Studio Assistant: Jeanine Jankovsky
Illustrator: Jim Swanson
Copy Editors: Barbara Webb, Dawn Kotapish
Technical Consultant: Joe Hansa
Intern: Kathryn Millhorn
With thanks to: Dave's Electric, Batavia, IL
    and Tech Lighting, Inc., Chicago, IL

Meredith® Books
Editor in Chief: James D. Blume
Design Director: Matt Strelecki
Managing Editor: Gregory H. Kayko
Executive Editor, Gardening and Home Improvement: Benjamin W. Allen

Director, Retail Sales and Marketing: Terry Unsworth
Director, Sales, Special Markets: Rita McMullen
Director, Sales, Premiums: Michael A. Peterson
Director, Sales, Retail: Tom Wierzbicki
Director, Book Marketing: Brad Elmitt
Director, Operations: George A. Susral
Director, Production: Douglas M. Johnston

Vice President, General Manager: Jamie L. Martin

Meredith® Publishing Group
President, Publishing Group: Stephen M. Lacy
Vice President, Finance & Administration: Max Runciman

Meredith® Corporation
Chairman and Chief Executive Officer: William T. Kerr

Chairman of the Executive Committee: E. T. Meredith III

The Home Depot®
Senior Vice President, Marketing and Communications: Dick Sullivan
Marketing Manager: Nathan Ehrlich

The editors of *Wiring 1-2-3®* are dedicated to providing accurate and helpful do-it-yourself information. We welcome your comments about improving this book and ideas for other books we might offer to home improvement enthusiasts.
Contact us by any of these methods:

1️⃣ Leave a voice message at **800/678-2093**

2️⃣ Write to **Meredith Books, Home Depot Books, 1716 Locust Street, Des Moines, IA 50309–3023**

3️⃣ Send e-mail to **hi123@mdp.com**. Visit The Home Depot website at **homedepot.com**

Canadian EDITION

Wiring 1-2-3

Install,
Upgrade,
Repair,
and
Maintain
Your Home's
Electrical
System

Meredith®
BOOKS

Wiring 1-2-3. TABLE OF CONTENTS

HOW TO USE THIS BOOK

Wiring 1-2-3 is filled with practical home wiring projects you can do! Professional electricians from The Home Depot stores across Canada and the United States provided projects that people like you want to do. They helped make sure the book has all the steps for you to successfully complete each project.

Start by reading **Safety First**. It walks you through basic rules for working with electricity. Pay close attention to bold red type and safety tips marked with stop signs.

If you're new to wiring projects, check out **Understanding Wiring**. It provides you with knowledge of what you're working with before you start a project.

To discover what electrical upgrades you may need in your home, turn to the chapter on **Inspecting Your Home**.

Ready to start a project? Check out **Top 10 Projects**. We've gathered projects The Home Depot electrical associates report are the most common for do-it-yourselfers.

Basic Tools and Skills will help you prepare for the projects in the following chapters: **Electrical Repairs, Planning Lighting,** and **Easy Upgrades**.

Electrical Repairs covers common repair jobs, such as rewiring lamps, replacing plugs and cords, and resetting breakers.

Good lighting does more than illuminate. Turn to **Planning Lighting** for help using lights to create indoor and outdoor areas with style and function.

Easy Upgrades shows you how to complete simple tasks that involve detaching old wires and attaching new wires. Projects include installing track lighting and grounding receptacles.

Once you've completed some easy upgrades or top 10 projects, you'll have the skills to move on to **Planning New Electrical Service**. Learn what tools you need and how to draw plans. Now you should be ready to tackle **Running New Cable, Installing New Services**, and **Major Projects**.

Running New Cable provides directions for installing lines in new framing and existing homes.

Turn to **Installing New Services** when you're ready to install a new electrical device or fixture such as a wall light, attic fan, or outdoor receptacle.

Major Projects covers installations that involve adding new circuits. Work through the projects in this chapter only after you've successfully completed projects throughout the book. If your local or provincial codes don't allow homeowners to add new circuits, this chapter will provide you with information to understand the project you're hiring an electrician to complete.

If you have a project involving telephone wire, coaxial cable, door chimes, or thermostats, turn to **Low-Voltage Wiring**.

If you prefer to hire a pro, this book will provide you with the knowledge to make the right choice and help you judge the work that's done.

For the do-it-yourselfer, Wiring 1-2-3 provides step-by-step directions, tips from the pros, lighting design ideas, and safety information to help you safely, easily, and accurately complete your home wiring projects and stylishly light your home.

TRICKS OF THE TRADE

Tips from the pros at The Home Depot® are scattered throughout this book.
Their expert advice will help you successfully complete the projects in *Wiring 1-2-3*.

SAFETY ALERT!
Prevent unsafe situations.

Homer's Hindsight
Avoid common mistakes.

A+ WORK SMARTER
Make smart work choices.

TRIP SAVER
Save time and mileage.

Designer Tip
Create a stylishly lit home.

OOPS!
Fix common mistakes. (Not that you'll make any.)

TOOL TIP
Use specialty tools to their best advantage.

BUYER'S GUIDE
Select the best materials.

GOOD IDEA
Info you need to know before you begin.

CLOSER LOOK
Understand all the details.

TIME SAVER
Save time and money.

SAFETY FIRST

Electricians and others who have done a lot of electrical work know they always have to follow safety precautions. They've heard hair-raising stories about what happens to people who ignore safe work habits. This book is loaded with safety reminders to help you stay safe while you work. Follow them.

Electricity deserves your respect. Consider how household current can affect the human body. If your feet are dry and you are wearing rubber-soled shoes, receiving a shock from a 120-volt circuit will definitely hurt, but it will probably not cause you serious harm. However, if conditions are wet or your feet are not protected with rubber-soled shoes and you are standing on the ground or on a metal ladder, 120 volts can cause the muscles in your hands to contract so that you grasp the source of current involuntarily. The current will cause your heart to beat wildly, very likely to the point of heart failure. Expect the same consequence if you touch both live wires of a 240-volt current, even if

SHUT OFF THE POWER. For most electrical projects, it is essential that you shut off power at the service panel. Flip a breaker off, or unscrew and remove a fuse.

TEST FOR POWER. Confirm that the power is off. (See pages 44–45 for how to use testers.) If you are using any type of neon tester, test the tester to be sure it is not showing power only because its bulb is out.

your feet are dry and protected. Children are in even greater danger.

The wiring in a modern home should have safety features, such as grounding and ground-fault circuit interruption. (See pages 51–57 for how to inspect your home for these and other safety considerations.) Both greatly reduce the possibility of dangerous shock, but they don't offer complete protection to a person who is working on exposed wires and devices. This is why professional electricians work very carefully. So should you.

Here are a few basic rules for safe electrical work. Follow them at all times, even when you are doing "just a little" electrical job.

SHUT OFF THE POWER

If there is no electrical current, you cannot receive a shock. Always shut off power to the circuit you are working on. Do this by flipping a circuit breaker or completely unscrewing a fuse. Then test for the presence of power (see pages 44–45).

TEST FOR POWER

Be aware that more than one circuit may be running in a box. Test all the wires in an open box for power, not just the wires you will be working on. Test everything twice.

Regularly test your tester to make sure it will indeed tell you when power is present. Touch it to a live circuit and see that it glows just before every test. Many a war story tells of someone turning off the power, only to have a family member or co-worker turn it back on while work is in progress. Post a sign telling people not to restore power; lock the service panel if possible.

STAY FOCUSED

Most electrical mishaps occur because of small mental mistakes. Remove all distractions. Keep people, especially children, well out of the way. Turn the radio off. Even after turning off the power, work as if the wires are live. Work methodically, and double-check all connections before restoring power.

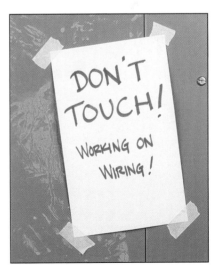

POST A SIGN ON THE PANEL. Take steps to make absolutely certain that no one will turn the power back on while you are working. If possible, lock the service panel.

SAFETY FIRST (CONTINUED)

USE PROTECTIVE TOOLS AND CLOTHING

Rubber grips offer more protection than simple plastic or wood handles, so always use rubber-gripped tools when wiring. Get in the habit of using them correctly: Grab by the handle, not the metal shaft. Make sure your pliers and cutting tools have rubber grips that are long enough so you will not be tempted to touch the metal while working.

Replace a tool when the rubber is damaged.

Wear rubber-soled shoes, such as athletic shoes, and perhaps rubber gloves, so that electrical current will not travel easily through you and into the ground. Never work with wet feet or while standing on a wet surface. Do not wear jewelry or a watch—anything that could possibly get snagged on wires. Use a fiberglass or wood ladder; an aluminum ladder conducts electricity.

ASK QUESTIONS

Electricians consult with each other all the time, even when they are 99 percent sure they already understand. Never proceed with an installation or repair unless you are completely sure you know what you are doing. Don't hesitate to ask "stupid" questions of electrical experts in a Home Depot or an electrical supply store.

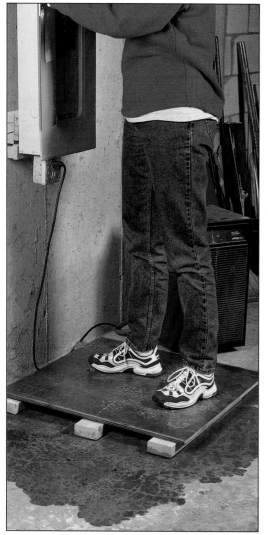

BE PROTECTED FROM THE GROUND UP. Always keep your feet protected with rubber soles, to lessen the effects of a possible shock. In damp areas, stand on dry boards.

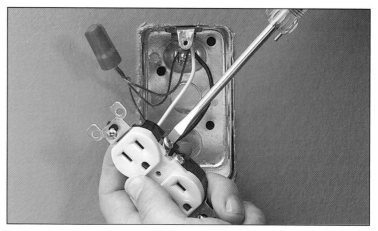

USE RUBBER-GRIPPED TOOLS. Don't use tools with plastic or wood handles unless they also have rubber sleeves to provide extra protection against electrical shock.

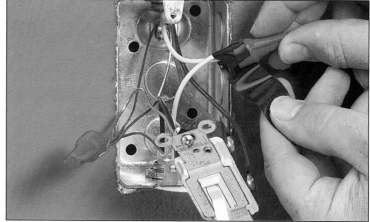

USE ELECTRICIAN'S TAPE. Electrician's tape provides extra protection against dangerous ground faults and shorts. Even if not required by code, wrap wire nuts and terminals with tape (see pages 43 and 50).

CHAPTER
1 UNDERSTANDING WIRING

Wiring contributes to the convenience of life. Flip a switch or turn a knob, and electricity instantly goes to work. Occasionally, however, a lamp flickers or a receptacle goes dead. Many electrical procedures are well within the range of most homeowners, but because electricity can be dangerous to work with, it may be tempting to call a professional electrician.

This chapter equips you with the basic knowledge you need to safely work on your home's wiring. It introduces you to the purpose and function of every wire and every device in your home. Take the time to become knowledgeable about how your home circuits work. Familiarize yourself with standard safeguards against electrical shocks and fire. For many projects, you may decide to call in a pro, but this chapter will prepare you to understand what makes a job safe and reliable—useful information whether or not you do it yourself.

CHAPTER ONE TOPICS

HOW POWER GETS DISTRIBUTED

Electricity is the flow of electrons through a **conductor**—copper or aluminum wires in household construction. Electricity must travel in a loop, called a **circuit.** In most cases, power travels out to a fixture or device through a hot wire—usually coated with black or red insulation—and back through a neutral wire, which has white insulation. When the circuit is broken at any point, power ceases to flow.

Newer homes are grounded. Grounding connects all outlets to the earth and is an essential safety feature (pages 11–13). Ungrounded outlets can give a serious shock if there is a short circuit due to a damaged wire or device. Ground wires are either bare copper or have green insulation. Polarization is a similar safety feature found in older homes (page 11).

VOLTAGE AND AMPS

Voltage is the electrical pressure exerted by the power source. Most household fixtures use 120 volts. Large items such as ranges and central air conditioners require 240 volts. On the packaging of electrical devices or in manufacturer's instructions, you may see voltage figures, such as 115, 125, 220, or even 250 volts. These numbers reflect the fact that voltage can vary; a 115-volt receptacle is interchangeable with a receptacle rated at 125 volts. **Here, we'll refer to 120-volt and 240-volt circuits.**

The pressure on all wires is approximately 120 or 240 volts, but the amount of electricity used by each fixture or appliances varies. This is because wires, fixtures, and appliances have different resistance to the voltage. Simply put, the thicker the wire, the more electricity travels through it. The terms **amperes** (or **amps**) and **watts** refer to the amount of electrical current that specific elements of a system use (pages 62–63).

FROM UTILITY TO HOME

Electrical power flows into neighborhoods through overhead (or underground) high-voltage wires. Transformers reduce the amount of power so that the wires entering homes carry a relatively safe load of 120 volts per wire. Through a **service head,** these wires enter a **meter.** The meter records how much power a home uses for billing purposes. The wires then enter the home's **service panel,** which divides the power into branch **circuits** (see opposite page).

Most homes have three wires— two "hot" wires to carry power into the house and one neutral wire to carry power back to complete the circuit. Having two hot wires means that a home can run **120-volt circuits** and **240-volt circuits** (for large appliances). Older homes with only two wires entering the home— a hot and a neutral—can run only 120-volt outlets. Some appliances, such as electric water heaters, are **hard-wired** to the circuit (wires are attached directly to the unit without the use of a plug or receptacle).

Underground electrical service to homes is usually provided by three wires—black, red, and white— that travel through a pipe called conduit. Overhead service is usually

CLOSER LOOK

TECHNICAL TERMS

If you need to plug in a lamp, you find an outlet, right? Well, almost. Technically, an electrical outlet is anyplace where electricity leaves the wires to perform a service— such as at a light fixture. A receptacle is a type of outlet; it is where electricity exits the system through, say, a toaster plug. A device is something that carries, but does not use, electricity itself; receptacles and switches are devices. A fixture is an electrical outlet that is permanently fixed in place, and an appliance is a user of electricity that can be moved. Thus an overhead light is a fixture, and a microwave oven is an appliance.

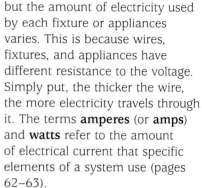

provided by three wires, one of which may be bare. They all connect to the house at a service head. These wires must not be damaged. If an overhead wire rubs against a tree or if an underground line seems exposed, call your utility company.

KNOW YOUR LIMITS

You can perform most repairs and installations on wires, devices, and fixtures in the home, but do not touch anything outside the home. Never touch wires leading to the service panel or wires upstream

from the main shutoff (pages 14–15). If you have questions about the wires entering your house or leading from the meter to the service panel, call your utility company. These wires are usually their legal responsibility.

FROM SERVICE HEAD TO RECEPTACLE

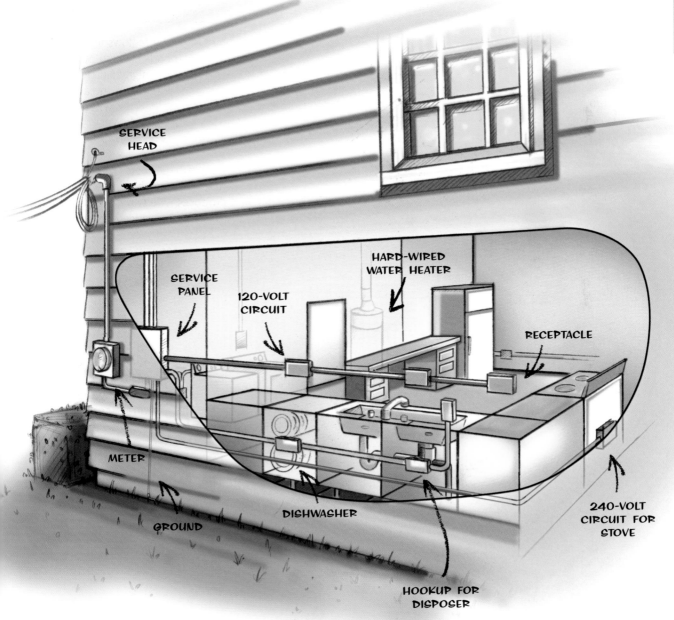

SERVICE HEAD

HARD-WIRED WATER HEATER

SERVICE PANEL

120-VOLT CIRCUIT

RECEPTACLE

METER

DISHWASHER

GROUND

240-VOLT CIRCUIT FOR STOVE

HOOKUP FOR DISPOSER

HOW A CIRCUIT WORKS

Service panels, whether they have breakers or fuses, divide household current into several circuits. Each circuit carries power from the service panel via hot (usually black or red) wires to various outlets in the house, and then back to the service panel via a neutral (usually white) wire.

TYPES OF CIRCUITS

Most household circuits carry 120 volts. There also may be several 240-volt circuits. Circuits are rated according to amps. If the outlets on a circuit draw too many amps, the circuit overloads. When this happens, a fuse will blow or a breaker will trip (pages 80–81), preventing an unsafe condition.

A 120-volt circuit usually serves a number of outlets. For instance, it may supply power to a series of lights, a series of receptacles, or some of each. Heavy-use items, such as dishwashers and refrigerators, must have their own dedicated circuits. A 240-volt circuit is always dedicated to one outlet. A standard 120-volt 15-amp circuit uses #14 wire; a 20-amp circuit uses thicker #12 wire. Until recently, 120/240-volt circuits used three wires—two hot and one neutral. Recent codes require four wires, as shown below; the added wire is for grounding.

Circuits provide convenience as well as safety. If a repair or new installation is under way, you can shut off power to an individual circuit instead of the entire house.

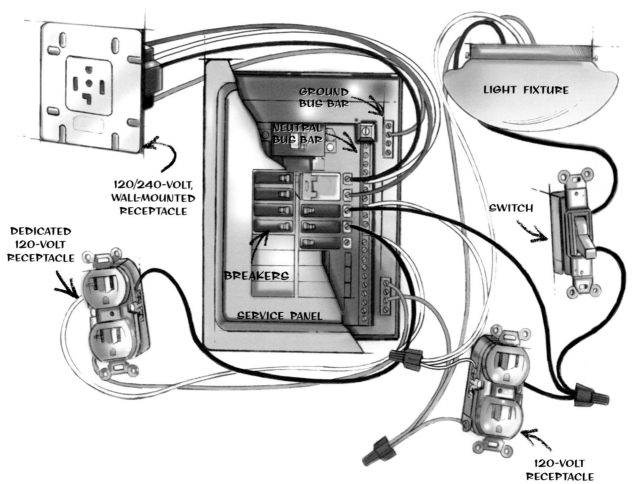

120/240-VOLT, WALL-MOUNTED RECEPTACLE

DEDICATED 120-VOLT RECEPTACLE

GROUND BUS BAR

NEUTRAL BUS BAR

BREAKERS

SERVICE PANEL

LIGHT FIXTURE

SWITCH

120-VOLT RECEPTACLE

A SERVICE PANEL HAS 120- AND 240-VOLT CIRCUITS. Your service panel distributes power according to the needs of a circuit. For example, a 240-volt circuit is designed to supply electricity to a heavy-duty user of power, such as an electric range or a dryer. The single receptacle on a dedicated 120-volt circuit might feed a refrigerator or a large microwave, while another 120-volt circuit feeds a series of receptacles and switched overhead light fixtures. **Depending on local code and the manufacturer, some switches may not have a grounding wire.**

GROUNDING AND POLARIZATION

Normally, electricity travels in insulated wires and exits through a light fixture into a bulb, for example. If a wire comes loose or if a device cracks, a short circuit **(ground fault)** results, energizing something you don't want to be energized. A short can occur if a loose wire inside a dryer touches the dryer's frame or if cracked insulation allows bare wire to touch a metal electrical box. If you touch energized metal, you'll get a dangerous shock. Grounding and polarization protect against this. Here's how they work:

GROUNDING

Grounding minimizes the possibility that a short circuit will cause a shock. A grounded device, fixture, or appliance is usually connected to a grounding wire—either bare or green—which leads to the neutral bar in the service panel. This bar is connected to the earth by one or a combination of these:

- cold-water pipe
- grounding rods driven deep in the ground
- metal plate sunk in a footing or in undisturbed soil.

When a ground fault occurs, the ground path carries the power to the service panel. This extra path lowers resistance, causing a great deal of power to flow back to the panel. This in turn trips a circuit breaker or blows a fuse. At the same time, power is directed harmlessly into the earth.

Whether your system uses grounding wires or conduit as the ground path, it must be unbroken in order to operate safely. A single disconnected ground wire or a loose connection in the sheathing or conduit can make the grounding system useless. To check whether a receptacle is grounded, plug in a receptacle analyzer (page 44).

POLARIZATION

Polarization is a way of making sure that electricity goes where you want it to go. Because a polarized plug has one prong wider than the other, there is only one way that the plug can be inserted into a polarized receptacle. If the receptacle is wired correctly and an appliance plug is polarized, the hot wire, and not the neutral wire, will always be controlled by the appliance switch. If the receptacle or plug isn't polarized, the neutral wire might be connected to the appliance switch instead, and power would be present in the appliance even when it is switched off. For extra protection against shock, install GFCI protection (page 27).

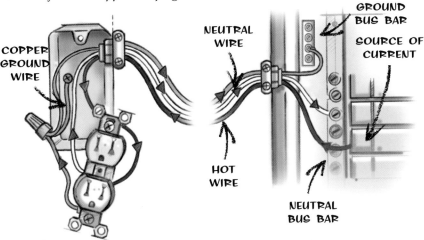

HOW A GROUNDED RECEPTACLE WORKS. To ground a receptacle, a ground wire (either bare or green) is attached to the receptacle (and to the box, if it is metal) and leads to the ground bus bar in the service panel. The panel itself is grounded (page 10). This receptacle is also polarized.

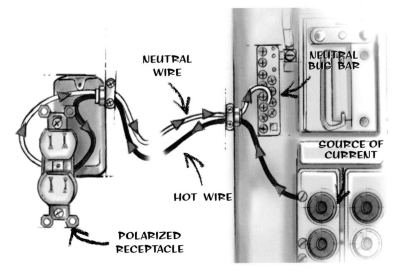

HOW A POLARIZED RECEPTACLE WORKS. The black wire is connected to the receptacle's brass terminal at one end and to the circuit breaker or fuse at the other end. The white wire runs from the silver terminal screw to the service panel's neutral bus bar.

GROUNDING METHODS

Before you begin any electrical work, find out how your system is grounded. First, plug a receptacle analyzer (page 44) into every receptacle to make sure they are grounded. Then check your service panel to make sure it's grounded—perhaps to a pipe or to grounding rods. Finally, look at the wiring of your receptacles or fixtures to see how they're grounded (opposite page).

If you have an older home without grounding, you should ground any new circuits you install. It is also possible to ground individual receptacles (page 107). A GFCI that is ungrounded can offer substantial protection for individual circuits (page 13).

JUMP A WATER METER. Provide an unbroken path for a ground that uses a water pipe. A water meter, for instance, breaks the path. Make sure that the ground wire is connected on the upstream side of the meter (toward the street, not the house) or that there is a tightly clamped jumper cable, as shown.

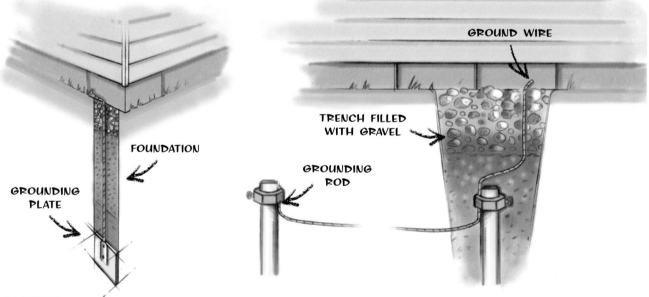

GROUNDING IN ROCKY SOIL. A house's ground wire may be attached to a grounding plate embedded in the concrete of a footing or foundation. To avoid subterranean rocks, the rod may be driven at an angle. Another solution is to connect the ground wire to rebar embedded in the foundation.

GROUNDING RODS. Usually, a standard grounding rod provides the best connection into the earth. Many systems use a single rod, but some codes require two connected rods, spaced at least 10 feet apart. A rod must be at least 10 feet long. Damp soil provides better grounding conditions than dry soil. If you have dry soil, add another rod or two to improve the connection. The grounding wire must be connected firmly to the rod, either with a special clamp or by welding. Local codes specify how deep the top of the rod should be buried.

HOW RECEPTACLES ARE GROUNDED

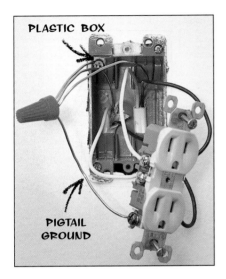

PLASTIC BOX

PIGTAIL
GROUND

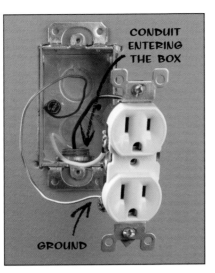

CONDUIT
ENTERING
THE BOX

GROUND

SAFETY
ALERT!

GROUND 'EM FIRST!
Always connect the ground wires first. Once you are sure all the ground connections are firm, connect the neutral wire, then the hot wire. To ensure a solid connection between the receptacle and the box, remove the cardboard washer from the receptacle's screws.
If you forget to ground a device, you may not detect the resulting danger because the ungrounded device or fixture will work just fine.

CABLES IN A PLASTIC BOX. Because plastic boxes do not conduct electricity, the receptacle must be grounded by attaching it to the bare ground wire in the cable. Check that bare copper grounding wires are spliced together and are attached to the grounding screw of the receptacle with a pigtail ground wire.

CONDUIT IN A METAL BOX. If conduit enters the box, it's likely it acts as the ground for the receptacle. To ground the receptacle, connect copper wire to the back of the box and the ground screw on the receptacle. Check the receptacle with an analyzer (page 44-45) to make sure it is grounded.

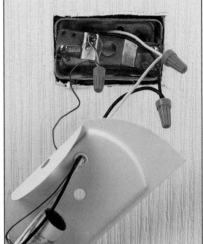

CABLE IN A METAL BOX. If one cable enters the box, you can ground the receptacle simply by attaching the ground wire directly to the ground screw on the receptacle and the box.

TWO CABLES IN A METAL BOX. If cable enters a metal box, it's likely the metal box is NOT grounded. Ground the receptacle the way you would a plastic box with cables coming into it. Use a wire nut to connect the two bare ground wires to a third wire that is attached to the receptacle's ground screw and to the box.

GROUNDING LIGHT FIXTURES. Your light fixtures should be grounded. Whether a light fixture box is metal or plastic, a grounding wire should be connected to the fixture. The ground connection may be made to a fixture's thin ground wire or to a screw on the mounting strap. If the box is metal, it should also be connected to a ground wire.

13

KNOW YOUR SERVICE PANEL

Find your home's service panel and learn how it works before you start wiring inspections, repairs, or installations. It's where you'll turn off power to circuits that you are working on and where you will run to when a circuit blows.

HOW A SERVICE PANEL WORKS

A service panel is the nerve center of your household's electrical system. It routes power to the circuits in your home and shuts down any circuit that gets overloaded. Every adult in your house should know how to safely reach the service panel to turn off or restore electricity.

Power from the utility company enters the panel through three thick main wires—two hots and one neutral. The main neutral wire connects to a neutral bus bar, and the two hot mains connect to the main power shutoff—either a large circuit breaker or a pull-out fuse.

Some panels use fuses (below left); some use breakers (below right).

Some very old homes have only two main wires, one hot and one neutral. Such a system is usually considered adequate if left alone, but if you add service to it you will be violating code. However, it will likely be inadequate for the electrical appliances in the average household. If this describes your home, get an electrician to install new service.

From the main shutoff, two hot bus bars (also called legs) run most of the length down the panel. Each bar carries 120 volts. Circuit breakers or fuses connect to these bars. (This is easier to see in a breaker box than in a fuse box.) Fuses and breakers rated at 120 volts are attached to a single hot bar; 240-volt breakers or fuses are attached to both hot bars.

Each 120-volt circuit has a black or color wire connected to a circuit breaker or fuse, and a white wire

NOTE: In some service panels you may find that the neutral feed wire is not white. If that's the case strips of white tape may have been added to differentate between the neutral and hot feeds.

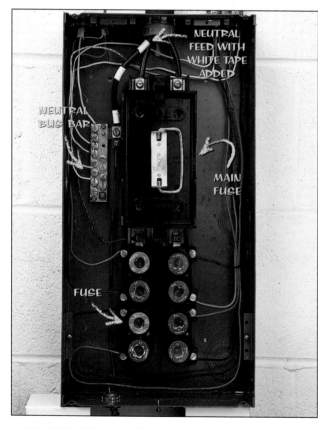

A 100-AMP FUSE BOX. You may find a 100-amp fuse box (above) instead of a breaker box as shown on the right. As long as the box is in good shape, and the connections are solid, the fuse box will do the job. If you are modifying an existing fuse box in any way it's better to replace it with a breaker box.

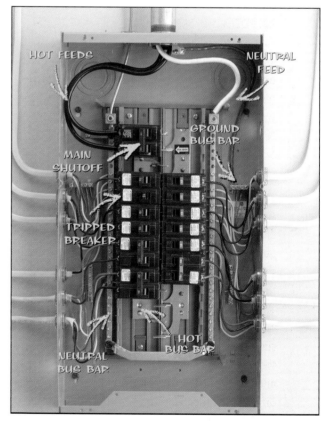

A 100-AMP BREAKER BOX. You can see the two hot bus bars and the ground bus bars more clearly on a breaker box. This 100-amp box has ample room for the wires, which are carefully laid out so you easily can see the path for each one.

connected to a neutral bus bar. Ground wires also lead to the neutral bar. Neutral bars (usually two) connect to a system ground wire. Some systems that use metal conduit or BX sheathing do not use grounding wires. (See page 13 for more about grounding.)

Power runs through each fuse or breaker and then out of the panel via a hot wire to whatever receptacles, lights, or appliances are on the circuit. White neutral wires bring power back to a neutral bus bar in the service panel, completing the loop.

WHEN CIRCUITS OVERLOAD

If a circuit becomes overloaded and is in danger of overheating, the circuit breaker will trip or the fuse will blow, disconnecting power to the entire circuit.

The same thing happens during a ground fault (page 11) or a short circuit, when a hot wire accidentally touches a neutral wire. *If a circuit shuts down frequently, you have a faulty appliance, device, or, most likely, an overloaded circuit (pages 62–63).*

WHEN TO UPGRADE A PANEL

If your panel seems cramped or confusing, have an electrician make sure it is safe. Some panels are too small for the number of circuits they serve, crowding the wires. Others have the neutral bars too near the hot bus bars so that hot and neutral wires are dangerously close to each other. Others have the neutral bar too far away, so neutral wires have to travel around the panel.

RESPECT YOUR SERVICE PANEL

Even seasoned electricians are very careful when working on service panels. If you have reservations about working on your service panel yourself, hire help. If you decide to inspect or work on the panel yourself, follow these safety tips.

■ **ALWAYS KNOW WHAT'S HOT.** A shutoff device—a switch, a breaker, or a fuse—turns off power only to the wires beyond the device. The wires entering the shutoff device are hot at all times. Be sure you know which wires are upstream of the shutoff (prior to the device and therefore not controlled by it) and which are downstream (after it, and therefore controlled by it). If you turn off the main breaker or remove the main

fuse, the whole house will go dead and all the circuits will be de-energized, but not all of the service panel will be safe. Unless the utility company turns them off, the thick wires entering the panel are always hot.

■ **KEEP YOUR HANDS OFF THE BUS BAR.** When you turn off a breaker or remove a fuse, the wires to the circuit will be dead, but the bus bar will still be hot. The bus bars are always energized unless the main breaker has been turned off or the main fuse has been removed.

■ **KEEP THE COVER ON.** Unless you are working on or inspecting a service panel, keep the cover attached so that there is no possibility that you will accidentally touch wires.

■ **MAKE A MAP OF YOUR CIRCUITS** (page 61) and post it on the inside panel door so you can easily see which breaker or fuse needs to be disabled.

■ **STORE STUFF AWAY FROM THE PANEL.** Keep flammable objects, including hanging clothes, at least 1 meter or 3 feet away. Have a charged flashlight handy.

■ **ALWAYS WEAR RUBBER-SOLED SHOES.** If the floor by the panel is at all damp, lay down some boards and lay a rubber mat on top of the boards.

■ **NEVER LET ANYONE CLIP TEMPORARY LINES INTO THE PANEL.** Welders and floor sanders sometimes want to clip 240-volt extension lines directly onto your hot and neutral bars. This is dangerous!

WIRES AND CABLES

U se the right wire and cable to avoid creating a dangerous situation that you'll have to tear out and redo. Here are the basics:

WIRES

Wire is usually made of a single, solid strand of metal encased in insulation. For flexibility and ease of pulling, some wire is stranded (above right). Wire is sized according to American Wire Gauge (AWG) categories. Size determines how much amperage the wire will carry. Common household wires and their ratings are:

- **#14 wire** (also called 14-gauge) carries 15 amps
- **#12 wire** carries 20 amps
- **#10 wire** carries 30 amps.

If a wire carries more amperage than it is rated for, it will dangerously overheat. Older wires have rubber insulation, which lasts about 30 years. New wires have longer-lasting polyvinyl insulation. Insulation color often tells the function of wire: **Black, red,** or other colors indicate hot wire. **White** or **off-white** generally is neutral. **Green** or **bare** wire is ground.

TYPES OF ELECTRICAL CABLE

Cable is two or more wires wrapped together in plastic or metal sheathing. **Nonmetallic (NM) cable** is permitted inside wall, ceiling, and floor cavities. Special metal plates must be added to the framing to protect the cable from puncture (page 129). Printing on **NM cable** tells you what is inside: 12/3 means there are three #12 wires, not counting the ground wire. "G" means that there is a ground wire. For underground installations and in damp areas, use **underground-feed (UF) cable (also called NMWU cable).** UF cable encases the wires in solid plastic. **Telephone cable** is being supplanted by **Cat 5 cable,** suitable for telephones, modems, and computer networking. **Coaxial cable** carries television signals. **Armored cable** (pages 124–125) has a flexible metal sheathing. **BX,** also called **AC90,** is a type of armored cable with a ground wire. **Metal-clad (MC) cable** has an insulated green grounding wire. (A similar material, Greenfield, or flexible conduit, is armored sheathing without wires. Install it, then pull wires through it.) **Conduit** is a solid pipe through which individual wires are run (pages 126–127). Metal conduit is often required in commercial installations. Most building departments require it only where the wiring is exposed. **Dedicated AC cable** is made especially for the run from the breaker box to the AC outlet.

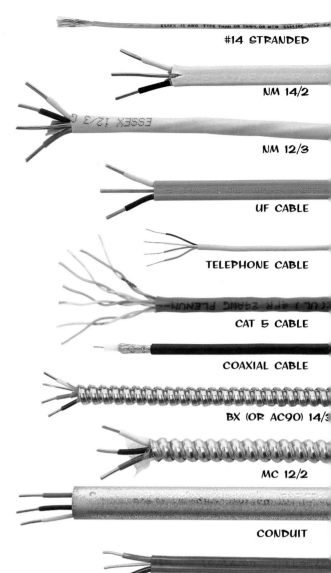

#14 STRANDED

NM 14/2

NM 12/3

UF CABLE

TELEPHONE CABLE

CAT 5 CABLE

COAXIAL CABLE

BX (OR AC90) 14/3

MC 12/2

CONDUIT

DEDICATED AC CABL

Homer's Hindsight

UPGRADE CABLE WHENEVER YOU CAN
While remodeling my old house, I pulled off the plaster and found cable running through the walls. It seemed in pretty good shape and had a ground wire, so I left it. Bad move. Electrical cable doesn't last forever. Even though the insulation wasn't cracked, chances are it will deteriorate within the next 20 years. I blew the chance to replace it easily.

RECEPTACLES AND SWITCHES

Switches and receptacles usually provide trouble-free operation for decades. However, they are not indestructible. If one of yours is cracked, singed, or seems too loose, replace it (pages 22, 26).

WIRES AND AMPS

Most switches and receptacles in a home are designed to carry 15 amps. To confirm yours, look on the metal plate for the amperage rating. Any 15-amp device should be connected to #14 wire (opposite page), which should lead to a 15-amp fuse or circuit breaker in the service panel.

If the wires are #12 or thicker, or if a 15-amp device is connected to a fuse or breaker that is 20-amp or greater, hire an electrician: This is a potentially dangerous situation.

Be sure that the amperage of a 240-volt receptacle is rated no lower than that of the appliance. If you are unsure as to which receptacle to use, check with your building department or ask an electrician.

GROUND HOLE UP OR DOWN?

In many areas, electricians install receptacles with the ground holes down (when the receptacles are vertical) or to the right (when they're horizontal). In other locales, the practice is just the opposite—up, or to the left.

Some professionals have complicated notions about why one way is better than the rest. Most of their theories involve making sure the ground prong is the last to be pulled out of a receptacle so the appliance remains grounded until it is plugged in. Ground down seems to be more common, but there is no harm in reversing this. Choose one way or the other, and then stick with it throughout your house.

CHOOSING A 120-VOLT RECEPTACLE

NEUTRAL SLOT

GROUND HOLE

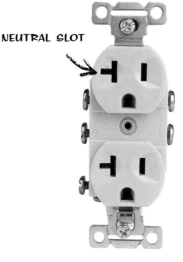

NEUTRAL SLOT

UNGROUNDED 120-VOLT RECEPTACLE. This type of receptacle has two slots, with no hole for a grounding prong. This one is polarized (page 11), with one slot longer than the other so that a polarized plug can be inserted only one way.

GROUNDED 15-AMP, 120-VOLT RECEPTACLE. This receptacle is the most common household electrical device. It will serve most lamps and appliances and will overload if you plug in two heavy-use items that total more than 15 amps.

20-AMP, 120-VOLT RECEPTACLE. This receptacle has a neutral slot shaped like a sideways T so you can confidently plug in large appliances or heavy-use tools. It must connect to #12 wires that lead to a 20-amp circuit or fuse in the service panel.

CHOOSING A 240-VOLT RECEPTACLE

WALL-MOUNTED 240-VOLT STOVE RECEPTACLE. Appliances using 240 volts have different plug designs so they can't be plugged into the wrong receptacle. To be safe, check the information plate on the appliance to confirm that the amperage matches that of the receptacle.

SURFACE-MOUNTED 120/240-VOLT RECEPTACLE. Some heavy-duty appliances require receptacles with both standard voltage and high voltage. For example, a range commonly uses 240 volts for its burners and 120 volts for the light and the clock. A 120/240-volt receptacle provides both levels of power.

WALL-MOUNTED 120/240-VOLT RECEPTACLE. This wall-mounted receptacle is typically used in a laundry room and installed in an appropriate electrical box.

CHOOSING A SWITCH

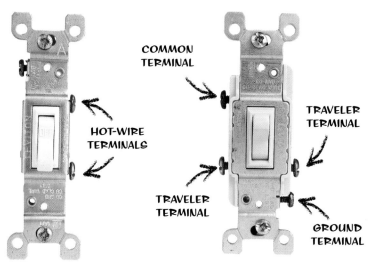

COMMON TERMINAL

HOT-WIRE TERMINALS

TRAVELER TERMINAL

TRAVELER TERMINAL

GROUND TERMINAL

SAFETY ALERT!

THREE OR FOUR?
Until recently, it was common to wire high-voltage receptacles with three wires—two hot and one neutral for a 120/240 receptacle; two hot and one ground wire for a 240-volt receptacle. Current codes, however, often require a fourth wire so that the receptacle has both a ground and a neutral for added protection (page 77).

SINGLE-POLE SWITCH. This is the workhorse switch in your home. It has two terminals for hot wires and may also have a green terminal for a ground wire. The toggle is labeled ON and OFF and should be connected to two #14 wires. These wires should be two black wires or a black wire and a white wire that has been marked (page 20). Check local codes.

THREE-WAY SWITCH. Three-ways are always installed in pair—both switches control the same light(s). There are no ON and OFF markings on the toggle. The common terminal is where you attach the wire bearing power from the source or to the fixture. (See page 24 for how to wire a three-way switch.)

RECEPTACLE AND SWITCH WIRING

Remove an electrical cover plate and pull out a switch or receptacle, and you'll find an arrangement involving a few wires going directly to the device. Or, you may find a multicolored tangle of wires, some related to the switch or receptacle and some not. Here are some of the most common wiring configurations you'll find behind electrical cover plates.

Switches that come with grounding terminals must be grounded to the system and the electrical box as seen in the photograph on the right. (For more information on grounding, see pages 12–13.)

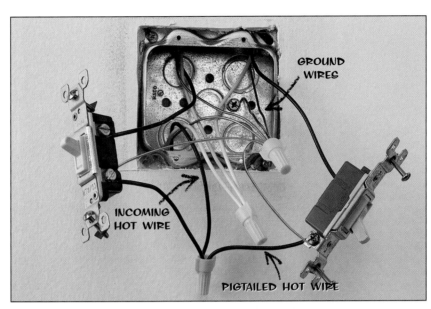

GROUND WIRES

INCOMING HOT WIRE

PIGTAILED HOT WIRE

SWITCHES SHARING A HOT WIRE. Switches that share a hot wire are on the same circuit. Two pigtails (page 49) branch off from the incoming hot wire and connect to each switch. Another hot wire runs from each switch to a light. White wires are spliced. If the switch has a ground terminal it must be grounded both to the system and the grounding terminal in the box.

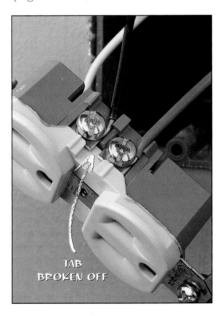

TAB BROKEN OFF

A SPLIT RECEPTACLE. Also known as a "half-hot" receptacle, this is connected to two hot wires. The brass tab joining the brass terminals has been broken off. With the tab broken, each hot wire energizes one plug. Some split-circuit receptacles have each plug energized by a different circuit so that you can plug in two high-amperage appliances without the danger of tripping a breaker. Others are wired so that half the receptacle is controlled by a wall switch, while the other half is hot all the time. See page 144 for more information.

CLOSE LOOK

MIDDLE-OF-THE-RUN RECEPTACLE

A receptacle with one cable that carries power into the receptacle and one that carries it to another device is called a "middle-of-the-run receptacle." Usually, two black wires are connected to the brass terminals and two white wires to the silver terminals. Or, the blacks and the whites may be joined, with a pigtail at each splice. Each pigtail is attached to the receptacle. If only one cable enters the box, the receptacle is at the end of the run. The black wire is attached to the brass terminal, the white wire is attached to the silver terminal, and the ground wire is attached to the receptacle.

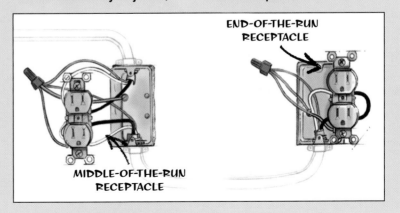

END-OF-THE-RUN RECEPTACLE

MIDDLE-OF-THE-RUN RECEPTACLE

RECEPTACLE AND SWITCH WIRING (CONTINUED)

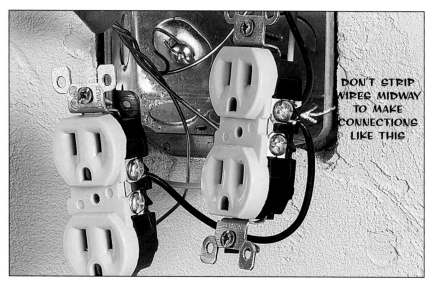

DON'T STRIP WIRES MIDWAY TO MAKE CONNECTIONS LIKE THIS

ALUMINUM WIRE IS SILVER IN COLOR

WIRES STRIPPED MIDWAY. You may find wires that have 1 inch of insulation stripped along the lengths, rather than being cut and each end stripped. Some electricians use this technique to save time. If the connections are tight, this is a safe arrangement. However, in most cases you should avoid this shortcut. Wire often gets nicked or scraped in the process. Use pigtails instead (page 49).

ALUMINUM WIRE. Aluminum wire, which is silver in color and thicker than copper wire, is not widely used because it expands and contracts, loosening connections. Make sure the receptacle is rated CO/ALR.(See page 64 for how to keep an aluminum system safe.)

SHOCK DEFENSE:
MAKE SURE ALL CIRCUITS TO THE DEVICE ARE OFF

Test for power (pages 44–45), or flip a light switch. Boxes may contain wires from more than one circuit. Trace the cables back to the service panel to turn off the circuit. Test all wires for power with a voltage detector (pages 5, 44, and 45).

HANDLE WITH CARE.
When removing the plate, grasp only the rubber handle of the screwdriver. When removing the device, pull gently, holding the plastic rather than metal parts. Don't dislodge wires. Wear rubber-soled shoes.

COLORED WIRES

WHAT COLORED WIRES MEAN. Colored wires are sometimes used by electricians to indicate different circuits. When this is done correctly, a circuit uses its own wire color—say, brown or purple. By turning off the breaker attached to the brown wire, you turn off power to all devices attached to brown wires. (Do not assume yours is wired this way. Always test to make sure power is off.)

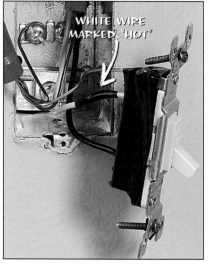

WHITE WIRE MARKED "HOT"

WHY WHITE WIRES MAY BE MARKED. When power runs into the fixture box rather than the switch box, another cable brings a black and a white wire into the switch box. When the switch is on, both wires are hot, so the white wire may be painted black with a marker or wrapped with a bit of electrician's tape. Do not remove the tape or scrape away the paint, or you will give the false—and dangerous—signal that the white wire is neutral.

2 TOP 10 PROJECTS

Home Depot associates across the United States and Canada report that the projects in this chapter are the most popular. You probably bought this book with one or more of these projects in mind. Some you might tackle out of necessity: One of your home's receptacles, switches, or light fixtures may be malfunctioning and therefore must be replaced. Other projects are upgrades that will make your home a safer and more enjoyable place to be.

Before beginning any of these projects, read Chapter 1, **Understanding Wiring,** for important background information. If you get into a project and find techniques that are new to you, follow the page references provided. They'll take you to the section that covers the subject in greater detail.

TOP 10 PROJECTS

CHAPTER TWO PROJECTS

REPLACING A SWITCH

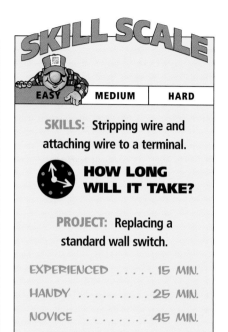

SKILL SCALE

EASY	MEDIUM	HARD

SKILLS: Stripping wire and attaching wire to a terminal.

HOW LONG WILL IT TAKE?

PROJECT: Replacing a standard wall switch.

EXPERIENCED . . . . . 15 MIN.

HANDY . . . . 25 MIN.

NOVICE . . . . 45 MIN.

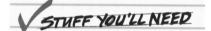

✓ STUFF YOU'LL NEED

TOOLS: Tester, combination stripper, lineman's pliers, longnose pliers, side-cutting pliers, screwdriver

MATERIALS: New switch, electrician's tape, wire nuts

BUYER'S 💲 GUIDE

STURDY DEVICES FOR HEAVY USE

If a switch is used constantly, pay a little extra for a device labeled "commercial" or "spec-rated." It has stronger contacts and is sturdier.

I f your switch pops when you turn it on, if it seems loose, or if your light fixture doesn't switch on even with a new bulb, it's time to replace the switch. Switches are easy and quick to test and install.

CHOOSING A REPLACEMENT

If the switch has two wires connected to it (it might also have a ground wire) and a toggle marked ON and OFF, it is a single-pole switch—the most common type.

If three wires connect to it (not counting the ground wire), it is a three-way switch (page 18).

If the box is not grounded you will have to test for power by placing one probe on a terminal and the other on a ground or neutral (white) wire. See page 13.

2 INSPECT THE WIRING. Remove the two screws holding the switch to the box. Gently pry out the switch. Pull on the wires to ensure that they're firmly connected to the terminals. If a wire is loose or broken, you've probably found the problem.

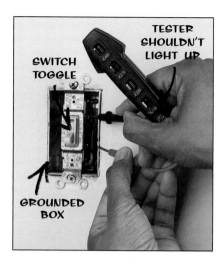

1 TESTING FOR POWER ON A GROUNDED BOX. Shut off the circuit breaker in the breaker box that supplies the switch. Remove the cover plate. Test for power with a 4-level voltage tester (also shown on page 44) or a Voltage Detector (page 45). Place one of the probes on a terminal screw and one on the box. The tester should not light up. Test the other side by switching the probe to the other terminal as well.

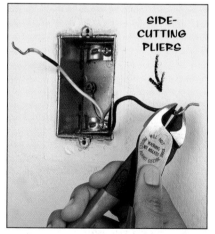

3 TRIM DAMAGED WIRE. Unscrew the terminal screws on the switch about ¼ inch (stop when they get hard to turn), and remove the wires. If a stripped wire end appears nicked or twisted, snip off the damage.

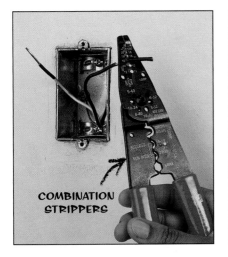

COMBINATION STRIPPERS

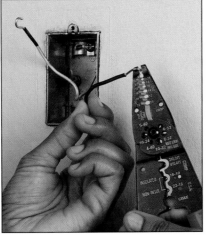

4 **RESTRIP TRIMMED WIRE.**
Using combination strippers, strip about ¾ inch of insulation from the end of any wires that you snipped (pages 46–47). If you strip a white wire that has been painted black or marked with black tape, remark it.

5 **TWIST A LOOP.** Form a question mark at the end of each wire, using the tip of combination strippers (page 46) or longnose pliers. Make the loop tight enough so that it just fits around the shank of the terminal screw.

Even if your local codes don't require that a switch or box be grounded, do it anyway. See page 11.

Homer's Hindsight

LOOP THOSE WIRES RIGHT THE FIRST TIME 'ROUND

Replacing a switch should be a no-brainer, shouldn't it? When my light fixture went on the fritz, I figured it had to be the switch. I replaced it but didn't loop the wires clockwise around the terminals. The fixture still didn't work, so I replaced it again. Still no light. Only then did I notice that a wire had spun off the switch terminal! I attached the wire the right way, and the light worked like a charm.

TOP 10 PROJECTS

LOOP WIRE CLOCKWISE ON TERMINAL

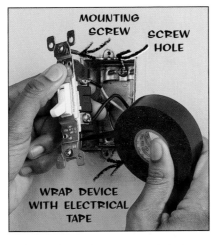

MOUNTING SCREW   SCREW HOLE

WRAP DEVICE WITH ELECTRICAL TAPE

6 **ATTACH THE WIRES.** On new switches, the terminals are screwed down tight. Unscrew each until it gets hard to turn. Slip a looped wire end under the screw head, with the end of the loop pointing *clockwise.* Squeeze the wire end tight around the terminal with longnose pliers (page 48) or the tip of a combination tool. Tighten the screw.

7 **WRAP THE SWITCH BODY.**
Although not required by code, for extra protection, wrap the switch with electrician's tape so that all terminals and bare wires are covered. This also ensures that wires won't come off the terminal screws.

8 **INSTALL THE SWITCH.** Gently push the wires back into the box as you push the switch back into position. Aim the mounting screws at the screw holes. Tighten the screws and check that the switch is plumb (straight up and down). The elongated holes allow for adjustments. Replace the cover plate, restore power, and try the switch.

REPLACING A THREE-WAY SWITCH

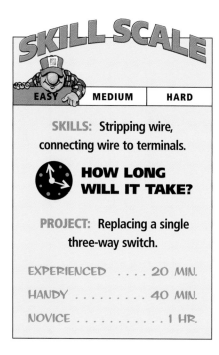

SKILL SCALE

| EASY | MEDIUM | HARD |

SKILLS: Stripping wire, connecting wire to terminals.

HOW LONG WILL IT TAKE?

PROJECT: Replacing a single three-way switch.

EXPERIENCED . . . . 20 MIN.

HANDY . . . . 40 MIN.

NOVICE . . . . 1 HR.

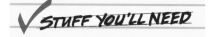

✓ STUFF YOU'LL NEED

TOOLS: Tester, screwdriver, combination strippers, longnose pliers, lineman's pliers

MATERIALS: Three-way switch, electrician's tape, masking tape

BUYER'S GUIDE

THREE-WAY DIMMER

Replace only one of your paired three-way switches with a dimmer: Two dimmers won't work. The remaining switch requires a standard three-way toggle. (Fluorescent fixtures also require special dimmers.)

Three-way switches work in pairs to control a light from two locations—handy for controlling a light from the top and the bottom of stairways, or from either end of hallways. The toggle isn't marked OFF and ON. Either up or down can be ON depending on the position of the toggle of the other three-way. (For more on three-way switches, see pages 148–150.)

Before you begin, shut off power to the circuit (page 5). Disconnect wires from terminals, and restrip any damaged wires (page 22). Most of the steps for replacing a three-way switch are the same as for a single-pole switch; but with three-ways, be sure to mark the wires before you remove the old device.

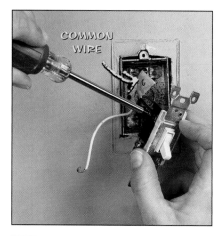

1 **TAG THE COMMON WIRE.**
Shut off power, remove the cover plate, and test to make sure there is no power in the box. Label the common wire with a piece of masking tape. The common terminal (page 149) is colored differently than the others (it's not the green ground screw) and may be marked "common" on the switch body.

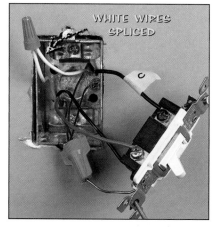

2 **OPTION A:**
WIRING ONE CABLE. When only one cable enters the box, it will have three wires plus a ground. Identify the hot wire using a voltage detector (pages 42, 45), or by touching one prong of a voltage tester to a ground and the other to each wire in turn. Attach the hot wire to the common terminal, which is a different color. Attach the other two wires to the traveler terminals. Connect the grounds.

OPTION B:
WIRING TWO CABLES. If two cables enter the box, one cable will have two wires and the other will have three wires (plus ground wires). But despite all the extra wires, you'll find only three wire ends. Proceed just as you would for a one-cable installation (left).

REPLACING A DIMMER SWITCH

✓ STUFF YOU'LL NEED

TOOLS: Screwdriver, side-cutting pliers, strippers

MATERIALS: Dimmer switch, wire nuts, electrician's tape

Make sure the new dimmer switch is rated for the total wattage of the fixture. A chandelier with eight 100-watt bulbs is too much for a 600-watt dimmer to handle. Don't use a standard dimmer for a fan, or you will burn out the motor. Install no more than one three-way dimmer; the other switch must be a three-way toggle. You can buy rotary dimmers (the least expensive), dimmers that look like standard switches, or models with a separate ON-OFF switch so the dimmer will turn on at the level of your choice.

1 **REMOVE THE KNOB.** Shut off power at the service panel. Pull off the rotary knob with firm outward pressure. Underneath is a standard switch cover plate. Remove the cover plate. Remove the mounting screws and carefully pull out the switch body.

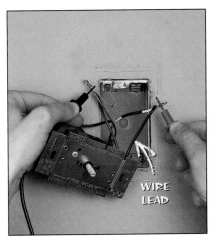

2 **TEST FOR POWER.** A dimmer has wire leads instead of terminals. Remove the wire nuts and test for power by touching the probes of the tester to both wires, or to either wire and the metal box, or to one wire and the ground wire. If power is detected, shut off the correct circuit in the service panel. (To test for continuity, see page 74.)

3 **OPTION A: INSTALLING A STANDARD DIMMER.** Attach the ground wire if there is one. Strip ¾ inch of insulation from each solid house wire and 1 inch from each stranded dimmer lead. Wrap a lead around a wire with your fingers so that the lead protrudes past the wire about ⅛ inch. Slip on a wire nut and twist until tight. Test the strength of the connection by tugging on both wires.

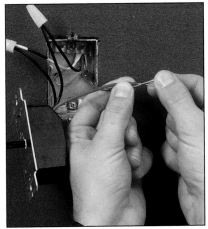

OPTION B: INSTALLING A THREE-WAY DIMMER. If you replace a three-way dimmer, tag the existing lead wires to connect the new dimmer in the same way as the old dimmer. If only one cable enters the box, attach the black wire to the common terminal and the other two wires to the traveler terminals. If you replace a three-way toggle switch with a dimmer, tag the wire that leads to the common terminal. The other two wires are interchangeable.

REPLACING A RECEPTACLE

TOP 10 PROJECTS

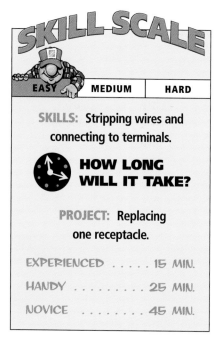

SKILL SCALE

| EASY | MEDIUM | HARD |

SKILLS: Stripping wires and connecting to terminals.

HOW LONG WILL IT TAKE?

PROJECT: Replacing one receptacle.

EXPERIENCED . . . . . 15 MIN.

HANDY . . . . 25 MIN.

NOVICE . . . . 45 MIN.

✓ STUFF YOU'LL NEED

TOOLS: Screwdriver, lineman's pliers, longnose pliers, side-cutting pliers, receptacle analyzer, combination strippers, level

MATERIALS: New receptacle, electrician's tape, wire nuts

If a receptacle doesn't seem to work, first check that whatever is plugged into it works properly. Replace any receptacle that is cracked. Before buying a replacement receptacle, check the wiring. Usually, the wires leading to a receptacle will be #14 and the circuit breaker or fuse will be 15 amp. In that case, install a 15-amp receptacle. Install a 20-amp receptacle only if the wires are #12 and the circuit breaker or fuse is 20 amps or greater.

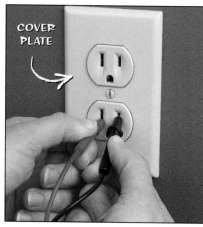

COVER PLATE

1 CHECK THAT POWER IS OFF. Turn off power to the circuit (page 5). Test to confirm. If the tester shows current, check your service panel and turn off another likely circuit. Test again, and proceed only if power is off. Remove the cover plate and unscrew the mounting screws. Being careful not to touch wires or terminals, pull out the receptacle.

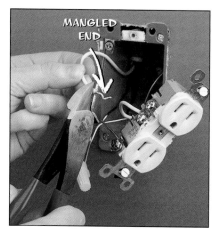

MANGLED END

3 SNIP AND RESTRIP DAMAGED WIRE ENDS. Once you're sure the power is off, unscrew the terminals and pull away the wires, taking care not to twist them too much. If a wire end appears nicked or damaged or if it looks like it's been twisted several times, snip off the end and restrip it (pages 22–23).

2 TEST WIRES FOR POWER. In a damaged receptacle, wires may be hot even though testing shows no power. Touch tester probes to the terminals. If more than two wires enter the box, test all the wires (page 5). If you have old wiring and both wires are black, use a receptacle analyzer (page 44) to check that the neutral wire is connected to the silver terminal and the hot wire to the brass.

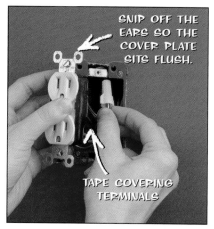

SNIP OFF THE EARS SO THE COVER PLATE SITS FLUSH.

TAPE COVERING TERMINALS

4 INSTALL THE RECEPTACLE. Wire the new receptacle the same as the old, with each white wire connected to a silver terminal and each black or color wire connected to a brass terminal. Wrap electrician's tape to cover all terminals and bare wires. Gently push the outlet into the box. Tighten the mounting screws, and check that the receptacle is straight. Replace the cover plate, restore power, and test with a receptacle analyzer.

ADDING GFCI PROTECTION

SKILL SCALE

| EASY | MEDIUM | HARD |

SKILLS: Stripping and splicing wires, connecting wires to terminals.

HOW LONG WILL IT TAKE?

PROJECT: Installing one GFCI receptacle.

EXPERIENCED . . . . 30 MIN.

HANDY . . . . 45 MIN.

NOVICE . . . . 1 HR.

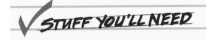

✓ STUFF YOU'LL NEED

TOOLS: Screwdriver, lineman's pliers, side-cutting pliers, combination strippers, level

MATERIALS: GFCI receptacle, electrician's tape, wire nuts

A ground-fault circuit interrupter (GFCI) shuts down power in milliseconds when it detects the tiniest change in current flow. Codes require GFCIs in bathrooms and outdoors. GFCIs are inexpensive and simple to install.

A single GFCI can protect up to four receptacles, switches, and lights on the same circuit. A GFCI circuit breaker can protect an entire circuit (page 100). If your home has ungrounded receptacles (pages 11–13), installing GFCIs will provide protection, but won't ground your circuits.

Check your GFCIs at least once a month by pushing in the test button. (The reset button should pop out. Push it back in.) A GFCI may provide power even though it has lost its ability to protect.

Don't use a GFCI as a receptacle for a refrigerator, freezer, or any other appliance that must stay on all the time—it may trip off without your knowing. Also, do not attempt to control a GFCI with a switch.

BUYER'S GUIDE

EXTEND A GFCI BOX

A bulky GFCI can dangerously crowd a box. Buy a box designed for raceway wiring and two 2-inch-long 6/32 screws. Fasten the screws through the GFCI and raceway box and into the box.

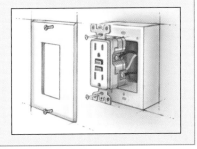

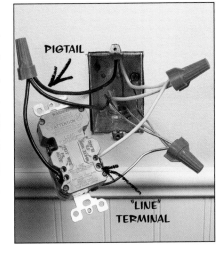

INSTALLING A SINGLE GFCI. Shut off the power. Make connections only to the LINE terminals. For an end-of-the-run box, connect the wires to the terminals. If the box is middle-of-the-run (shown), for each connection, make a pigtail by stripping either end of a 6-inch-long wire. Splice each pigtail to the wire(s) with a wire nut, and connect it to the GFCI terminals. Put the white wire on the silver terminal and the black or color wire on the brass terminal.

PROTECTING OTHER OUTLETS. Shut off the power. Connect the wires carrying power into the box to the LINE terminals. Then connect the wires leading out of the box (to other receptacles or lights) to the LOAD terminals. If you're unsure which wires come from the service panel, pull the wires out of the box and position them so they will not touch each other, restore power, and use a tester to see which pair of wires is hot; connect these to the LINE terminals.

TOP 10 PROJECTS

UPGRADING A CEILING FIXTURE

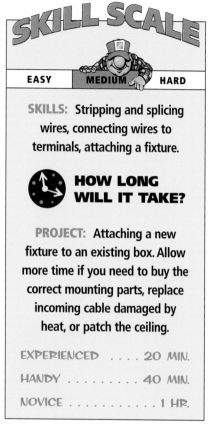

SKILL SCALE

EASY   MEDIUM   HARD

SKILLS: Stripping and splicing wires, connecting wires to terminals, attaching a fixture.

HOW LONG WILL IT TAKE?

PROJECT: Attaching a new fixture to an existing box. Allow more time if you need to buy the correct mounting parts, replace incoming cable damaged by heat, or patch the ceiling.

EXPERIENCED . . . . 20 MIN.

HANDY . . . . 40 MIN.

NOVICE . . . . 1 HR.

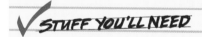

STUFF YOU'LL NEED

TOOLS: Voltage tester or multitester, combination strippers, lineman's pliers, longnose pliers, side-cutting pliers, screwdriver, stepladder

MATERIALS: New light fixture, electrician's tape, wire nuts

Before you replace the fixture, check that the canopy of the new fixture (the part that snugs up to the ceiling) will cover any imperfections in the drywall or plaster. If you have a thin "pancake" box, replace it with a remodeling box (pages 133–134).

Determine which mounting hardware you'll need before buying a new fixture. Turn off the power at the service panel (page 5) and remove the fixture. Enlist a helper to support the fixture while you remove the mounting screws that hold the canopy in place. Gently pull down the fixture. Working as if the wires are hot, unscrew the wire nuts. Test that the power is off, and then undo the wires. Note the type of mounting hardware, or remove it and take it along when buying a new fixture.

The new fixture will probably include mounting hardware (usually, a strap). You may be able to reuse existing hardware.

Always push the house wires up into the box. Never place them in the fixture's canopy, where they may be harmed by heat.

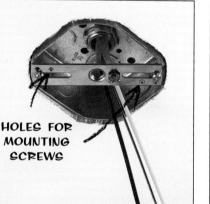

HOLES FOR MOUNTING SCREWS

1   **WIRE A FLUSH-MOUNTED FIXTURE.** Tug on the hardware to make sure the box is firmly attached. Don't depend on the wires and wire nuts to support it while you work. Rest the fixture on a stepladder, or make a hook from a wire coat hanger and temporarily suspend the fixture from the mounting strap. With the power off, splice white to white wires and black to black wires, using wire nuts (page 47). Tuck the house wires up into the box.
Note: Sometimes you will find that two black wires come with the fan or fixture. If this is the case, the smooth wire is hot and the ribbed wire will be the neutral.

2   **MOUNT THE FIXTURE.** Slide a mounting screw through the fixture and up into the threaded hole in the strap. Start one mounting screw, fastening it halfway in, then start the other screw. With a screwdriver or a drill and screwdriver bit, drive the mounting screws tight.

SAVE THAT INSULATION
Don't remove the fiberglass insulation at the top of a ceiling fixture, even if it seems to get in the way. It's there to protect the wires from overheating.

SPECIAL ALIGNMENT

SWIVEL STRAP

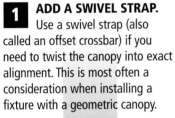

1 **ADD A SWIVEL STRAP.** Use a swivel strap (also called an offset crossbar) if you need to twist the canopy into exact alignment. This is most often a consideration when installing a fixture with a geometric canopy.

MOUNTING BOLT

MOUNTING NUT

2 **MOUNT A FIXTURE THAT NEEDS ALIGNMENT.** Wire the fixture, and tuck the house wires up into the box. Screw both mounting bolts into the threaded holes of the strap. Line up the fixture-mounting holes with the bolts and attach the fixture, using the decorative mounting nuts.

CENTER-MOUNTED

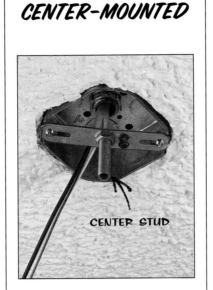

CENTER STUD

1 **USE A CENTER STUD FOR CENTER-MOUNTED FIXTURES.** A center stud, sometimes called a nipple, may have wires running through it (for a pendant fixture) or around it (for a center-mounted fixture). A variation uses a center nipple, which screws into the strap.

MOUNTING NUT

2 **INSTALL A CENTER-MOUNTED FIXTURE.** The center stud should be long enough to go through the strap and the fixture, but not so long that it pokes into the box or hits the globe. Attach the wires and fold them into the box as you slide the canopy up and over the stud. Snugly secure the canopy with the nut.

OLDER INSTALLATION

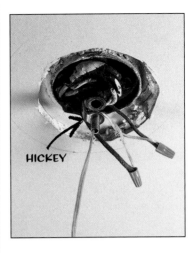

HICKEY

1 **ADD A HICKEY IN OLDER INSTALLATIONS.** An older home may have a $\frac{3}{8}$-inch pipe running through the middle of the ceiling box. To install a pendant fixture, add a hickey to make the transition from the pipe to a new fixture. Feed the fixture leads through the hickey, and tuck the house wires up into the box.

2 **ADD A MEDALLION.** Hickeys are found in old houses with plaster and lath ceilings. The fixture canopy usually won't cover damaged plaster around the ceiling box. Adding a medallion saves you the trouble of patching and painting while adding a decorative feature.

TOP 10 PROJECTS

ADDING UNDER-CABINET LIGHTS

| EASY | MEDIUM | HARD |

SKILLS: Running new cable and connecting to power, attaching light fixtures, stripping and splicing wires, wiring a switch.

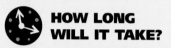

HOW LONG WILL IT TAKE?

PROJECT: Removing the backsplash, cutting holes, running cable, wiring four lights, and reinstalling the backsplash.

EXPERIENCED . . . . . . 1 DAY

HANDY . . . . 1.5 DAYS

NOVICE . . . . 2 DAYS

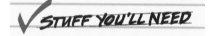

TOOLS: Multitester or voltage tester, lineman's pliers, side-cutting pliers, longnose pliers, combination strippers, drywall saw, screwdriver, drill with spade bit, utility knife, flat pry bar

MATERIALS: Under-cabinet fluorescent lights, armored or NM cable, electrician's tape, wire nuts, cable clamps, cable staples, switch box, nailing plates

Under-cabinet lighting brightens work surfaces adding focus and depth to counters. Fluorescent lighting is a cool, low-energy light source and can be dimmed with a special dimmer. If you need only one or two lights, consider fixtures complete with cords and switches. For a longer series of lights controlled by a wall switch, mount individual fixtures and run cable to a powered receptacle (if local code allows) or pull power from an approved source. Get the longest lights possible for the available under-cabinet spaces to ensure even coverage.

Some codes require that you run armored cable or conduit through walls; others allow exposed armored or nonmetallic (NM) cable.

In this project involving an existing receptacle, cables are run and the fixtures are installed as shown below, then the wall switch is wired. At this point you will turn off the power to the existing receptacle and wire the lights into it. Restore power and test.

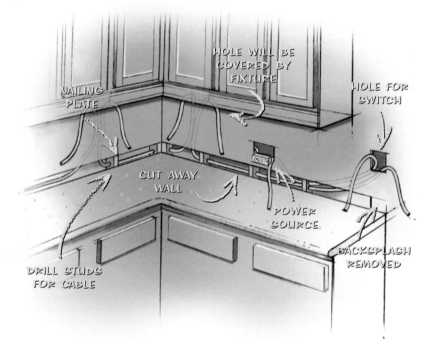

HOLE WILL BE COVERED BY FIXTURE

NAILING PLATE

HOLE FOR SWITCH

CUT AWAY WALL

POWER SOURCE

DRILL STUDS FOR CABLE

BACKSPLASH REMOVED

1 RUN CABLE. Plan the wiring so as many holes as possible will be covered when you're done. If the countertop backsplash is removable, remove it and cut a channel in the drywall or plaster that will be completely covered by the backsplash. Drill holes in the studs to accommodate cable (pages 128–132). (If you can't remove a backsplash, allow time for patching and painting the wall afterward. Or, install tile between the countertop and the wall cabinets.) Examine each light to determine exactly where the cable will enter and exit. Cut narrow holes in the wall where the cable will enter the lights.

Cut carefully so the hole will be covered when the light is installed. Cut a hole for the switch box, and run cable into it from a power source—perhaps a nearby receptacle (pages 141–142). Do not connect the cable to power. Run cable from the switch box to the hole for the first light, then from the first to the second light, and so on. Let about 16 inches of cable hang from the holes so you'll have plenty of slack to make connections. Most local codes allow fluorescent lights to be used as junction boxes, so you can string the wire from light to light. Check to be sure.

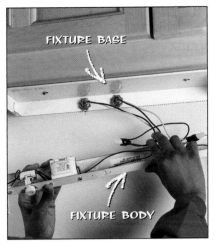

FIXTURE BASE

FIXTURE BODY

2 **ATTACH THE LIGHTS.** Disassemble the lights, and remove the lens and fluorescent tubes. Clamp each cable to the light as you would clamp cable to a box (pages 123 and 125). Have a helper hold the light as close to the rear wall as possible while you drive screws through the light and into the underside of the cabinet. Be sure that the screws won't poke through to the inside of the cabinet.

3 **WIRE THE LIGHTS.** Plan so that wires will not come within an inch of the ballast. Splice wires with the leads inside the light, black to black and white to white. Position the wires flat in the base so they will not get in the way when you add the fixture body. Gently push the bottom portion of the light into position. If it does not go in easily, take it down and realign the wires for an easier fit. Attach the fixture base.

Designer Tip

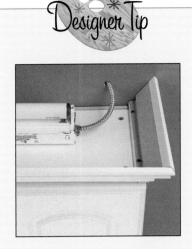

ADD COVE LIGHTING
If your cabinets have space above, you can install lights there without hiding the cable.

4 **WIRE THE SWITCH.** Install a switch box (pages 122–123). Splice the white wires together. Attach each of the black wires to a single-pole switch. (If you wish to dim the light, use a dimmer made especially for fluorescent fixtures.) Connect the ground wire to the switch and to the box if it is metal. Cover the terminals with tape. Shut off power to the receptacle or junction box that will supply the power. Splice white to white and black to a black or color wire in the receptacle (page 47). Restore power.

TIME SAVER

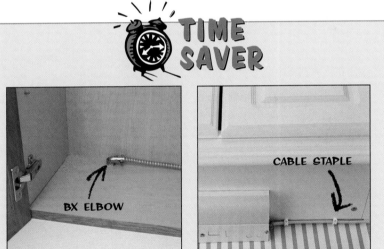

BX ELBOW

CABLE STAPLE

CABLE INSIDE THE CABINET
Cut holes in the cabinets. Lay BX or MC cable on the inside. Plan exactly where the cable will enter each light below the cabinet. Because you can't slip excess cable into the wall cabinet, you'll have to cut the cable precisely. (See pages 124–125 about working with armored cable.)

CABLE UNDER THE CABINET
Attach the lights under the cabinets, string cable under the cabinet, and staple the cable in place using cable staples. Measure and cut carefully so the cable is flat along the length. Check your local code before doing this; it is not allowed in some areas.

HANGING A CEILING FAN

TOP 10 PROJECTS

SKILL SCALE

EASY	MEDIUM	HARD

SKILLS: Removing an old box, installing a fan box, attaching a fixture, stripping and splicing wires, wiring a switch.

 **HOW LONG WILL IT TAKE?**

PROJECT: Replacing a light fixture with a new ceiling fan.

EXPERIENCED . . . . . 2 HRS.

HANDY . . . . 4 HRS.

NOVICE . . . . 8 HRS.

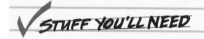
STUFF YOU'LL NEED

TOOLS: Drywall saw, hammer, voltage tester or multitester, combination strippers, linesman's pliers, longnose pliers, crescent wrench, side-cutting pliers, screwdriver, reciprocating saw or metal-cutting keyhole saw

MATERIALS: Fan, fan-rated box, downrod extender, light kit, fan/light switch (can be remote-controlled), electrician's tape, wire nuts

Ceiling fans circulate air downward to cool rooms in the summer and upward to evenly disperse heat in the winter. Observe the following guidelines to install a fan, and it will effectively circulate the air in your home without hissing, wobbling, or pulling away from the ceiling.

PLANNING FOR A FAN

Before installing a fan, consider these issues:

- Decide whether to wire the switch to control the fan and the light separately (page 34).
- Buy a separate light kit if your unit doesn't include one; some fans include lights, so check to be sure.
- Plan how you'll cover the hole once you remove the old ceiling box. Buy a light with a canopy that's wide enough to cover the hole, or get a medallion to hide ceiling imperfections (page 29).
- Decide how many blades you want. Four-blade fans move more air than five-blade models.
- Avoid "ceiling hugger" fans—they do not circulate air well. Fans should have downrods long enough (you can buy downrod extenders) to position fan blades at least 10 inches from the ceiling, but check that the blades are no lower than 7 feet from the floor.
- Use only a fan-rated dimmer if you install one; a standard dimmer will burn out the fan motor.

REMOVING AND REPLACING THE BOX

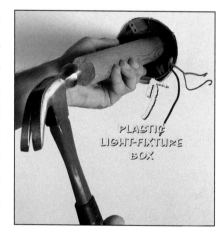

PLASTIC LIGHT-FIXTURE BOX

FAN-RATED PANCAKE BOX

REMOVING A CEILING BOX. Shut off the power. Sometimes, you can remove screws or nails and pry out the box. Or, you may have to carefully cut away drywall or plaster to get to fasteners. If the box is nailed to a joist, cut around the box to enlarge the hole, and tap the box loose using a piece of wood and a hammer. You may be able to cut through fasteners with a reciprocating saw or a metal-cutting keyhole saw. Take great care not to slice through any cable.

SCREWING A FAN BOX TO A JOIST FROM BELOW. If you have a joist in the middle of the hole (as may be the case if you removed a thin "pancake" ceiling box), attaching a fan box from below will be easy. Buy a thin fan-rated box, and clamp the cable to it. Hold it in place and drill pilot holes; then drive in 2-inch wood screws. (Don't use drywall screws or "all-purpose" screws—they break too easily.)

ADDING A NEW CEILING BOX

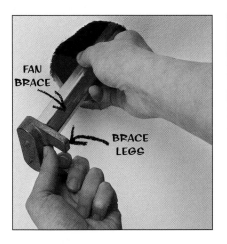

FAN BRACE

BRACE LEGS

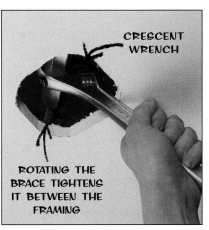

CRESCENT WRENCH

ROTATING THE BRACE TIGHTENS IT BETWEEN THE FRAMING

SAFETY ALERT!

1 **SLIP IN THE BRACE.** Check the joists for any wiring or plumbing runs that might be in the way before you install the brace. Test-fit the box on the brace; then take it apart again. Push the brace through the hole and spread it until it touches the joists on both sides with the legs of the brace resting on top of the drywall or plaster.

2 **TIGHTEN THE BRACE.** Measure to make sure that the brace is centered in the hole. Position it on the joists at the correct height so the box will be flush with the surface of the ceiling. Use a crescent wrench or channel-type pliers to tighten the brace until it is firm.

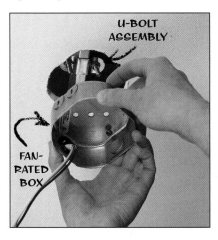

U-BOLT ASSEMBLY

FAN-RATED BOX

A+ WORK SMARTER

3 **ATTACH THE BOX.** Attach the U-bolt assembly to the brace so that the assembly is centered in the hole and the bolts face down. Thread cable through the cable connector and into box. Slip the box up so the bolts slide through it, and tighten the nuts to secure the box.

Check for obstructions such as plumbing pipes or wiring runs that may be attached to the ceiling joists before you install the fan brace.

WHEN FRAMING IS ACCESSIBLE, ATTACH A CEILING BOX TO A JOIST.
Install this type of box in unfinished ceilings or ceilings with a large hole. Drill pilot holes and drive in 1¼-inch wood screws to attach it to a joist.

OR INSTALL A BRACED BOX FROM ABOVE.
Buy a new-work ceiling fan box with a brace. Slide the box along the brace to position it. Tighten the clamp. Attach the brace by driving in 1¼-inch wood screws.

CEILING JOIST BOX

BRACED BOX

Fiberglas insulation

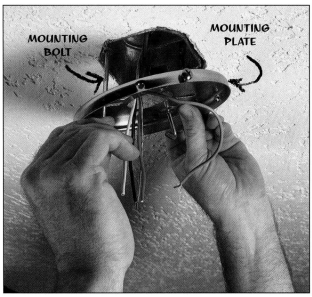

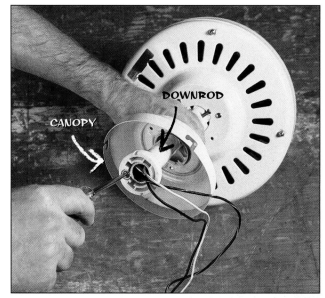

4 **INSTALL THE MOUNTING PLATE.** Thread the wires through the center of the mounting plate. If the box has mounting bolts that poke through the plate, fit the mounting plate over the bolts and fasten it with the nuts provided. If separate bolts are provided, push each one through the mounting plate as shown. When both bolts are in place, tighten the plate onto the ceiling.

5 **ASSEMBLE THE DOWNROD AND CANOPY.** On a work table, ready the fan for installation, following manufacturer's instructions. Run the fan leads through the downrod (or downrod extender), and tightly screw on the downrod. Remember to tighten the setscrews. Slip on the canopy, then install the bulb-shape fitting at the top of the downrod. It will rest in the canopy when the canopy is attached to the ceiling. Be careful not to mangle the wires. Do not attach the fan blades yet.

CLOSER LOOK

SWITCHING THE FAN AND LIGHT
You probably have two-wire cable (not counting the ground wire) running into the ceiling fixture. If so, you have four options to control the fan and the light:

■ Hook the fan to the two wires so the wall switch turns the fan and the light on or off at the same time. Use the pull chains on the fixture to control the fan and light individually. This is convenient enough if you don't need to change fan speeds often.

■ Purchase a fan that has a special fan/light switch that requires only two wires. These fans are expensive, however, and the switches have been known to turn the fan on by themselves — a dangerous situation if you're away for a few days.

■ Install a remote-control switch, as shown on page 35. This is rather costly, but far less work than running new cable.

■ Run three-wire cable from the fixture to the switch and hook it up as shown on page 22. You can conveniently control the fan and light separately by using a special wall switch.

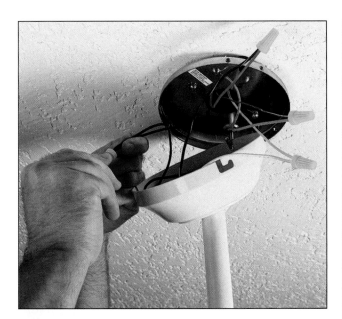

6 **WIRE THE FAN.** Temporarily hang the fan from the hook on the mounting ring. Connect the copper ground wire to the green wire attached to the fan base. If you have only two wires, connect both the black lead (for the fan motor) and the blue or striped lead (for the light) to the black house wire, and the white lead to the white house wire. If you have three-wire cable, connect black to black, white to white, and red to the blue or striped light lead. Check the manufacturer's directions. You may choose to install a remote control unit (right).

7 **ATTACH THE CANOPY TO THE MOUNTING PLATE.** Use a helper to support the fan motor while you drive the screws. Push the wires and wire nuts up into the box to keep them from vibrating against the canopy when the fan is running. Clip the canopy onto the mounting plate and tighten the screws.

BUYER'S GUIDE

REMOTE SWITCH

HANGING BRACKET

RECEIVING UNIT

WIRELESS REMOTE SWITCH

If you have only two wires running from the switch to the fan box, a remote control will let you control the fan and light separately.

Before you install the canopy, hook up the receiving unit with both fan and light leads (black and blue or striped) spliced with the remote's black lead, and the white wire spliced to the white lead. Make sure the little dip switches are set the same on the switch unit and the receiving unit. Install the canopy. Put a battery in the sending unit, and attach a hanging bracket on a wall.

If the ceiling fixture was originally switched, the two wires sending power to the fan are still controlled by that switch. The sending unit controls the fan or light, or both, only when the wall switch is on.

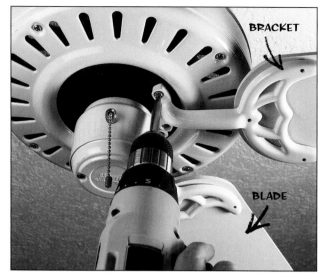

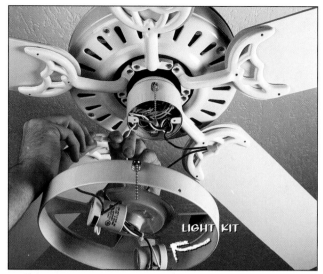

8 **ATTACH THE BLADES.** If the brackets are not all of uniform shape, return them and get replacements. Screw a bracket to each fan blade. Make sure the side of the blade that you want to show faces down. Attach each bracket to the motor with two screws. Drive the screws slowly to avoid stripping. Don't bend the brackets as you work.

9 **WIRE THE LIGHT KIT.** When you remove the plate on the bottom of the fan, you may see a tangle of wires. Don't worry; just find the blue or striped lead and the white lead, and connect them to the light kit leads. Screw the light kit up to the fan. Some light kits require a spacer ring between the fan and the light. The spacer ring should come with the kit or the fan.

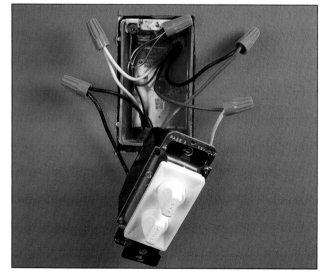

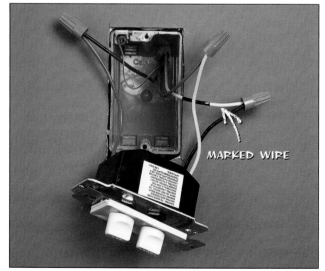

10 **OPTION A:**
IF POWER RUNS TO THE SWITCH. Shut off power to the switch. A three-wire cable usually runs from the fan to the switch, and a two-wire cable brings power to the switch. Follow the manufacturer's instructions; wire colors vary. Most likely you'll splice the black wire bringing power to the black switch lead, and splice the two white wires together. Then splice the black wire from the fan and the red wire from the light to the switch's fan and light leads.

OPTION B:
IF POWER RUNS TO THE CEILING BOX. If the switch box has only one cable—the one from the fan—then power runs to the ceiling box. Usually you'll find a black-marked white wire that brings power from the fan to the switch; however, the previous installer may not have marked it. If unmarked, wrap tape around its end and splice it with the switch's black lead. Splice the red wire (from the light) and the black wire (from the fan) to the switch's light and fan leads.

INSTALLING RECESSED LIGHTING

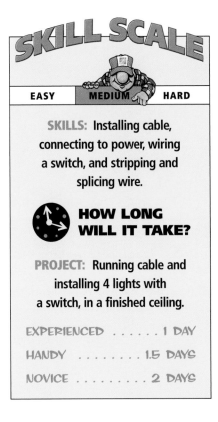

SKILL SCALE

EASY   MEDIUM   HARD

SKILLS: Installing cable, connecting to power, wiring a switch, and stripping and splicing wire.

HOW LONG WILL IT TAKE?

PROJECT: Running cable and installing 4 lights with a switch, in a finished ceiling.

EXPERIENCED . . . . . . 1 DAY

HANDY . . . . 1.5 DAYS

NOVICE . . . . 2 DAYS

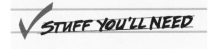

STUFF YOU'LL NEED

TOOLS: Stud finder, drill with long bit, drywall saw or hole-cutting drill attachment, voltage tester or multitester, combination strippers, lineman's pliers, screwdriver, safety glasses

MATERIALS: Can lights and trims, switch box and switch, cable and clamps, electrician's tape, wire nuts

an lights, also called "pot lights," are recessed lights that use 60- to 150-watt floodlight bulbs. They're ideal for task lighting, highlighting artwork, or grouped to illuminate whole rooms. (See page 93 for tips on planning.) Cans get hot. Position them at least 1 inch away from wood and other flammables. Always follow manufacturer's instructions.

If the joists are exposed, use a new-work can light (page 39). For ceilings already covered by drywall or plaster and lath, buy a remodel can (below) that clips into a hole cut in the ceiling, (It's also called an old-work, or retrofit, can.) To install a remodel can, follow the steps beginning on page 38.

CHOOSING CANISTER LIGHTS

Can lights are designed to suit specific situations. Here's how to choose the right one:

- If there's insulation in the ceiling, buy IC (Insulation Compatible) lights. Standard recessed lights will dangerously overheat when surrounded with insulation.

- Buy tiny low-voltage can lights. They're stylish accent lights, but they are expensive. They're wired in the same way as standard can lights.

- Use bulbs of the recommended wattage or lower. Bulbs with too-high wattage will dangerously overheat. When putting a number of cans on a dimmer, add up all the wattage and make sure your dimmer is rated to handle the load.

- If you have less than 8 inches of vertical space above the ceiling, purchase a low-clearance canister.

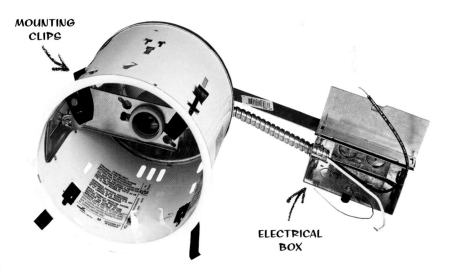

MOUNTING CLIPS

ELECTRICAL BOX

ANATOMY OF A CAN LIGHT. A standard remodel canister fixture has an approved electrical box, suspended far enough from the light so it will not overheat. A thermal protector shuts the light off if it becomes too hot (for example, if you use a bulb of too-high voltage). If you have less than 8 inches of vertical space above your ceiling, purchase special cans designed to fit into this smaller space. Be sure they are IC (Insulation Compatible) rated so there will be no danger of overheating.

INSTALLING RECESSED LIGHTING (CONTINUED)

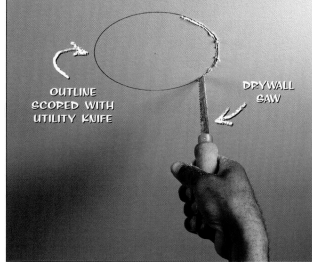

OUTLINE SCORED WITH UTILITY KNIFE

DRYWALL SAW

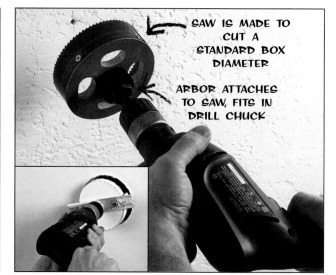

SAW IS MADE TO CUT A STANDARD BOX DIAMETER

ARBOR ATTACHES TO SAW, FITS IN DRILL CHUCK

1 **OPTION A:**
CUT THE HOLE. Lightly mark all light locations. Use a stud finder to make sure they do not overlap a joist. Or, drill a hole and poke a bent wire up into it to make sure the hole is entirely between joists. Use the template provided with the light to draw a circle on the ceiling. Draw and cut each hole precisely. If it is even a little too big, the can may not clamp tightly. Wearing safety glasses, cut the line lightly with a utility knife; then cut along the inside of the knife line with a drywall saw. Take care not to snag any wires that may be in the ceiling cavity.

OPTION B:
USE A HOLE-CUTTING SAW. This tool saves time and cuts holes precisely. You don't have to draw the outline of the hole on the ceiling; just mark the center point. Check to see that you will not run into a joist. Check that the lights fit snugly without having to be forced into place. Note: This tool is costly (the saw and the arbor are sold separately), but it's worth the price if you have more than six holes to cut through plaster. A less-expensive tool (inset) is available for cutting through drywall only.

16 INCHES OF EXTRA CABLE

Designer Tip

OPEN

FISH-EYE

BAFFLE

REFLECTIVE

CHOOSING CANISTER LIGHTS
Most can lights have two parts—the body and the trim. Choose both at the same time, or install bodies that have many trim options, and choose later. Open trim is the simplest and least expensive option. Baffle trim diffuses light so it is more evenly distributed. Fish-eye (also called eyeball) trim swivels to highlight a decorative feature. Reflective trim offers maximum brightness.

2 **ROUGH-IN THE WIRING.** Run cable from a power source to a switch box, and then to the first hole, allowing at least 16 inches of extra cable to make wiring easy. (See pages 122–131 for how to run cable.) Work carefully and use a drill with a long bit to avoid cutting additional access holes (pages 133–138) that will need patching later.

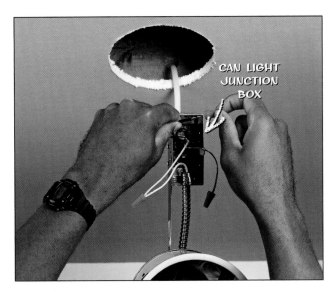

3 **WIRE THE LIGHT.** Open the light's junction box. Usually, there's a plate that pops off. Run cable into the box and clamp it. Strip insulation and make wire splices—black to black, white to white, and ground to ground (pages 46–47). Fold the wires into the box and replace the cover.

4 **MOUNT THE LIGHT.** Most remodel cans have four clips that clamp the can to the ceiling by pushing down on the top of the drywall or plaster. Pull the clips in so they do not protrude outside the can. Slip the can's box into the hole; then push the can body up into the hole until its flange is tight to the ceiling. With your thumb or a screwdriver, push each clip up and outward until it clicks and clamps the fixture.

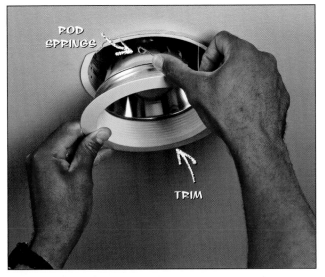

CLOSER LOOK

5 **ADD THE TRIM.** Most trims are mounted with coil springs or squeezable rod springs (as shown). If you have coil springs: Hook each spring to its assigned hole inside the can (if it is not already there). Pull out each spring and hook it to the trim; then carefully guide the trim into position. If you have rod springs, squeeze and insert both ends of each spring into their assigned holes; then push the trim up.

MOUNTING A NEW-WORK CAN LIGHT

If ceiling joists are exposed, this is an easy installation. On a workbench, attach the plaster ring to the fixture. Adjust it to compensate for the thickness of the ceiling drywall that will be installed later. At the ceiling, slide the mounting bars outward so they reach joists on each side. Hammer the four tabs into the joists. Add 1¼-inch screws for extra strength.

INSTALLING LOW-VOLTAGE LANDSCAPE LIGHTING

SKILL SCALE

| EASY | MEDIUM | HARD |

SKILLS: No special skills are required.

HOW LONG WILL IT TAKE?

PROJECT: Installing a series of 4 to 6 outdoor lights.

EXPERIENCED . . . . . . 2 HR.

HANDY . . . . 3 HR.

NOVICE . . . . 4 HR.

✓ STUFF YOU'LL NEED

TOOLS: Screwdriver, drill, tool for trenching, lineman's pliers

MATERIALS: Set of lights with a transformer and timer

MAKE THE LAYOUT AND CONNECTIONS. Attach the transformer/timer to the wall near a receptacle. Lay the lights on the ground where you want them, and string cable alongside them. When you are satisfied with the layout, make the snap-on electrical connection for each light; squeeze each gently with pliers to ensure a tight fit. Poke each light into the ground. If a light won't push in, don't bang on it; instead, slice the ground with a shovel or flat pry bar, then poke into the slice. Dig a shallow trench for the cable, or cover it with mulch.

WIRE AND PROGRAM THE TRANSFORMER. Thread the cable through the clamp, and poke the wire ends into the terminals. Tighten the setscrews and the clamp. Some transformers allow you to control two or more sets of lights. If there is a HI/LOW switch, set it to HI if the cable extends very far; see the manufacturer's instructions. Program the timer according to the directions.

Installing outdoor lights is simple and quick. And low-voltage cable carries so little power that there is no danger of shock. The most economical approach is to buy a set that includes lights, cable, and transformer/timer. Avoid the very cheapest models, which have unreliable connections.

Make sure the cable is long enough for your needs. If you need to extend a line beyond the recommended length, use #14 or #12 low-voltage wire from beginning to end. The system plugs into a standard 120-volt receptacle. If you don't have an outdoor receptacle, you can plug the transformer into an indoor receptacle and run the low-voltage cable out through a small hole in the wall. (To install an outdoor receptacle, see pages 162–163.) See pages 94–96 for planning outdoor lighting.

3 BASIC TOOLS AND SKILLS

Equipped with a basic understanding of household electricity, you may be tempted to dive right into your project. After all, how hard can it be? Grab a utility knife, some tape, and a pair of pliers, and start splicing and twisting, right?

Some homeowners who tackle wiring projects with this attitude successfully complete the repairs they set out to make. But there's no guarantee that their work will meet the standard requirements for safety and longevity.

Professional electricians perform their highly detailed work accurately and safely. They ensure the tightness of connections so there's no chance of them coming apart. They cover all bare wires to avoid the danger of shorts.

With the help of this book, the right tools, and some practice, you can maintain and upgrade your home's electrical system with the confidence and reliability that rival the pros'.

CHAPTER THREE TOPICS

THE RIGHT STUFF
Although you'll need carpentry tools to cut and patch holes for installing cable and boxes, don't use them as substitutes for tools designed specifically for electrical work. The right tools protect you from shocks, result in secure splices and connections, and make the job more enjoyable.

BASIC TOOL KIT

Compared to power tools used for carpentry work, the cost of electrical tools is a drop in the tool bucket. Buy everything you need. If you spend a little more to buy professional-quality tools, you'll find that they'll help you work faster and produce better connections.

The following section describes the tools you'll need to make all the inspections, repairs, and installations described in this book through page 108. (You may also need a few basic household tools such as a hammer, standard pliers, and a keyhole saw.) More advanced tools required for installing new electrical lines are described on pages 110–111. Be sure all your metal tools have insulated grips.

TOOLS YOU'LL NEED

With **combination strippers,** you can remove insulation from wires neatly and without nicking the metal. This tool is far superior to cheap adjustable strippers, which are a struggle to use and often damage the wire.

A **wire stripper/cutter** cuts wire like a pair of scissors and has a hole for stripping wire. Professional electricians often use this tool, or lineman's pliers, to strip wires instead of using combination strippers. It takes practice to do this without damaging the wire. **Side-cutting pliers,** or diagonal cutters, make it easy to cut wire and to snip off stripped plastic sheathing.

With **lineman's pliers** you can cut wire and easily twist them together. Buy a high-quality pair that is fairly heavy in the hand, smooth-operating, with precisely aligned cutting edges for easy snipping of wires. Use **longnose pliers** to twist a tight loop in a wire end before attaching it to a terminal. Make sure the one you buy is sturdy enough to handle household wiring—some are intended for finer wires used in electronics.

Among the many precautions you can take to protect against electrical shock, using **rubber-grip**

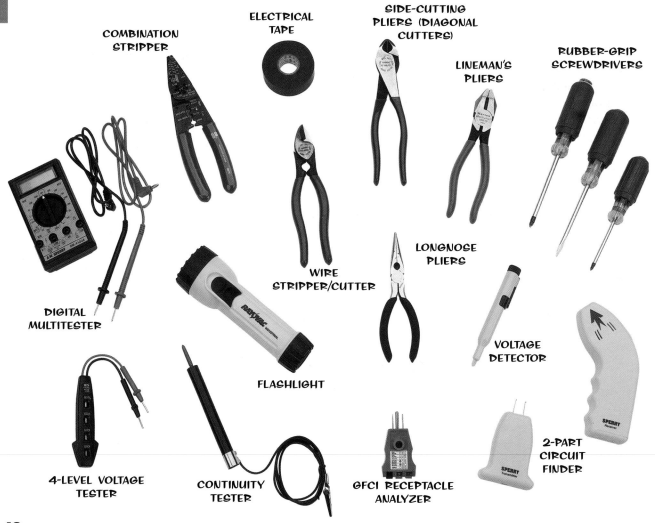

COMBINATION STRIPPER

ELECTRICAL TAPE

SIDE-CUTTING PLIERS (DIAGONAL CUTTERS)

LINEMAN'S PLIERS

RUBBER-GRIP SCREWDRIVERS

DIGITAL MULTITESTER

WIRE STRIPPER/CUTTER

LONGNOSE PLIERS

FLASHLIGHT

VOLTAGE DETECTOR

2-PART CIRCUIT FINDER

4-LEVEL VOLTAGE TESTER

CONTINUITY TESTER

GFCI RECEPTACLE ANALYZER

screwdrivers when doing electrical work is one of the most important. Don't use screwdrivers with plastic handles only. They can crack, creating a shock hazard. The handles should be large enough so that you will not be tempted to grab the metal shaft while you work. (Four-in-one screwdrivers are especially unsuited to electrical work because they have a metal shaft that runs through the handle.)

Keep a reliable **flashlight** handy because you may have to work in the dark.

Every home center has a bin of inexpensive **electrical tape.** It'll do the job, but far better is the more expensive, better-quality tape—it's thicker, more adhesive, and longer lasting.

SELECTING TESTERS

Even if you do not plan to do much electrical work, buy a **GFCI (Ground-Fault Circuit Interrupter) receptacle analyzer** (it handles standard receptacles as well). It will quickly tell you whether the receptacles are safe.

There are a variety of tools you can use to test for the existence of power. A **continuity tester** checks the reliability of fuses, switches, and sockets with the power off. A **four-level voltage tester**—better than the cheaper, single-level version—indicates if the power is on or off. A **digital multitester** is useful for appliance repair as well as electrical work. It performs the tasks of both a continuity tester and a voltage tester.

A **voltage detector** senses power, even through wire and cable insulation, so you can see whether wires are live before you work with them. With a **two-part circuit finder,** you can easily find out which circuit a receptacle is on. (Testers are described in detail on pages 44–45.)

BASIC TOOL CARE

Protect tools from moisture; rust causes them to lose their effectiveness. Make sure that the plastic insulation on each tool is in good shape so that your hand does not touch any metal part. If a cutting tool loses its edge so it's a struggle to cut wire, replace the tool.

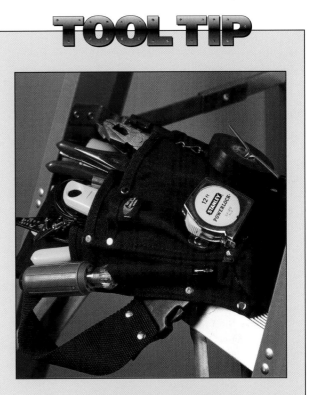

TOOL TIP

ELECTRICIAN'S TOOL BELT
Though a basic carpentry tool belt will keep your electrical tools close at hand, an electrician's tool belt is specially designed for keeping often-used electrical items within easy reach. Even if you work on only a half-dozen boxes and devices, a belt will save time.

FIBERGLASS STEPLADDER
Never stand on a metal ladder while working with or near electricity. Use a fiberglass ladder, like the one shown above, or a wood ladder that's labeled "nonconductive." Not only do these items protect you from shock, they are also heavier than aluminum ladders and are more stable.

WORK SMARTER

SQUARE DRIVE SCREWS
Slotted screws can be difficult to set, especially in tight spaces. Square drive screws (with a square slot in the head) require a special head for the screwdriver but offer a more positive driving action with far less slippage.

USING TESTERS

Reliable testing is essential to electrical work. Testers tell you whether the wires you are working on are hot; whether a switch, receptacle, or fixture is in working order; and whether a receptacle is wired safely.

Don't skimp on electrical testing tools. You might not need a fancy multitester, but avoid inexpensive tools such as single neon testers. They can quickly burn out or easily break, making you think there is no power when there really is.

If you buy a multitester, invest in a digital model rather than one with a dial. A digital tester is easier to use and is far less likely to give the wrong reading.

Examine your tester regularly to be sure it provides accurate information. To confirm that it's working, poke the probes of a voltage tester into a receptacle you know to be live, and make sure the tester lights up. Touch the probes of a continuity tester together. If the tester lights up, it's working. If it doesn't light up, it may need a battery or a bulb. Keep testers dry and safe from harm.

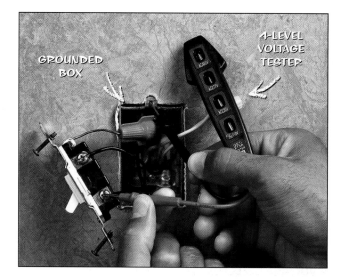

A VOLTAGE TESTER INDICATES THE PRESENCE OF POWER. A four-level voltage tester is safer and more reliable than one-level versions. Always confirm that a voltage tester is working by trying it on a circuit that you know to be live. Touch the tester's probes to a hot wire and a grounded box, to a hot wire and a neutral wire, or insert them into the slots of a receptacle. If the tester light doesn't come on, the circuit is shut off.

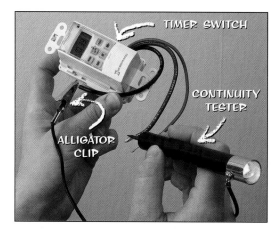

A CONTINUITY TESTER TELLS YOU WHETHER A DEVICE OR FUSE IS DEFECTIVE. Disconnect the device from all household wires. Attach the tester's alligator clip to one terminal and touch the probe to the other terminal. If the device switch is working, the tester light will glow when the switch is turned on and go out when the switch is turned off. To test the wiring in an appliance or lamp, touch both ends of each wire. The tester light will glow if the wire is unbroken. (To test a fuse, see page 81.)

A RECEPTACLE ANALYZER TELLS YOU WHETHER YOUR RECEPTACLES ARE SAFE. When you plug this analyzer into a receptacle, one or more of three lights will glow, telling you whether the receptacle is working, grounded, and polarized (page 11). Red analyzers will test ground-fault circuit interrupter (GFCI) receptacles as well as standard receptacles. Yellow analyzers test standard receptacles only.

VOLTAGE DETECTOR

A VOLTAGE DETECTOR SENSES POWER—EVEN THROUGH WIRE AND CABLE INSULATION. This handy tester lets you check whether wires are live before you work on them. The probe doesn't need to touch a bare wire or terminal. Press the detector button and hold it on or near an insulated wire or cable to see if power is present. If there's power, a light comes on.

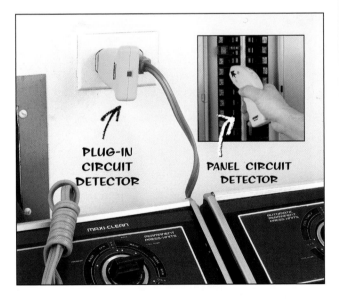

PLUG-IN CIRCUIT DETECTOR

PANEL CIRCUIT DETECTOR

CIRCUIT DETECTORS INDICATE WHICH CIRCUIT A RECEPTACLE IS ON. Plug one part of the circuit detector into the receptacle. Open the service panel door and point the other part of the tester at the circuit breakers. The detector will glow to indicate the correct circuit. Even after switching off the circuit, check for power at the receptacle before working.

TOOL TIP

MULTITESTERS

SET TO VOLTS

TO TEST FOR VOLTAGE. A multitester tests 120-volt, 240-volt, or low-voltage circuitry. Multitesters have negative and positive probes. Test for voltage by touching each probe to a wire, terminal, or receptacle slot. You can also touch one probe to the black wire and the other to a ground, such as a metal box. The display should show between 108 and 132 volts for a 120-volt circuit, and between 216 and 264 volts for a 240-volt circuit. Low-voltage circuitry can register as low as 4 volts.

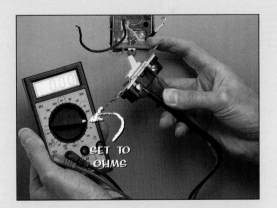

SET TO OHMS

TO TEST FOR CONTINUITY. A multitester can test a switch, receptacle, fuse, or light fixture to see whether its circuitry is damaged. Shut off the power and remove the device. Test for continuity by turning the dial to an "ohms" setting and touching each probe to a terminal on the device. If you test a switch, turn it ON. If the multitester needle shows zero resistance, the device is in good shape. An infinity reading means that the device is defective.

STRIPPING AND SPLICING WIRE

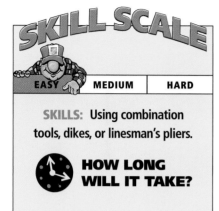

SKILL SCALE

EASY	MEDIUM	HARD

SKILLS: Using combination tools, dikes, or linesman's pliers.

HOW LONG WILL IT TAKE?

PROJECT: Stripping and splicing two wires.

EXPERIENCED . . . . . . 1 MIN.

HANDY . . . . 3 MIN.

NOVICE . . . . 10 MIN.

With practice, you'll soon learn to remove insulation and connect wires with ease. Keep in mind that cutting into metal wire while stripping will weaken it. If wires are not joined tightly, the electrical connection will be compromised and could cause a short.

To work with new cable, you'll first have to remove the sheathing. (See pages 122–125 to see how to remove sheathing from various types of cable.)

Use a wire nut (page 50) to join wires. Twist the wires together before adding the wire nut.

Homer's Hindsight

SPLICING WIRES

A guy I know makes his splices without twisting the wires together. He just holds the wire ends next to each other and twists on a wire nut. "Just as strong," he said. So I did it his way. Most of the splices were OK, but one came loose, shutting down a whole circuit. It took me hours to find it. Next time I'll twist the wires together first.

STUFF YOU'LL NEED

TOOLS: Combination strippers, lineman's pliers, or side-cutting pliers

MATERIALS: Wire, electrician's tape, wire nuts

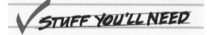

When splicing two wires together, strip about 1 inch of insulation. If the wire will be joined to a terminal, remove about 3/4 inch.

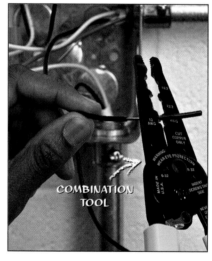

COMBINATION TOOL

1 **OPTION A:**
STRIP WIRES WITH A COMBINATION TOOL. Don't use a utility knife; it will probably nick the wire. Choose a pair of wire strippers you are comfortable with, and practice with them until you're comfortable using them. To use a combination tool, slip the wire into the correct hole, squeeze, twist, and pull off the insulation. The insulation should come off easily.

STRIPPER/CUTTER

OPTION B:
USE A WIRE STRIPPER/CUTTER. Many electricians consider combination strippers too slow. They prefer tools that are sometimes called "dikes." These include lineman's pliers, side-cutting pliers, or a stripper/cutter, with a single stripping hole. It takes time to learn to use these tools without nicking the wire. Press down with just the right amount of pressure to cut through the insulation and not the wire. Maintain the same pressure and twist until the insulation is cut all the way around. Ease up on the squeezing pressure, and pull off the insulation.

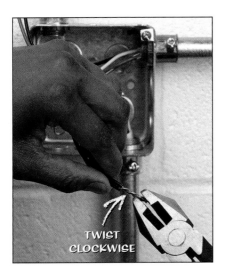

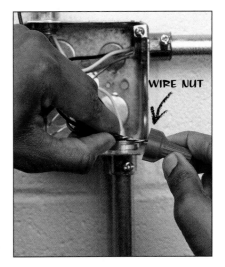

2 **TWIST WIRES TOGETHER.**
Hold the stripped wires side by side. Grab the ends of both with linesman's pliers. Twist clockwise, making sure that both wires turn. Twist them together like a candy cane; don't twist one around the other. The wires should form a neat-looking spiral. Twist several times, but don't overtwist or you might break the wires.

3 **CUT THE END.** Using the linemann's pliers or side-cutting pliers, snip off the end of the twist. Leave enough exposed metal so that the wire nut will just cover it—about ½ inch usually does it.

4 **CAP WITH A WIRE NUT.** Select a wire nut designed for the number and size of wires you have spliced (page 50). Slip the nut on as far as it will go, and twist clockwise until tight. Test the connection by tugging on the nut—it should hold securely for extra protection. Wrap electrician's tape around the bottom of the cap.

A+ WORK SMARTER

SPLICING THREE OR FOUR WIRES.
When twisting three or four wires together, hold them parallel and twist them all at once with linesman's pliers. (Don't twist two together and then try to add a third.) Choose a wire nut designed to accommodate the number and size of wires you have spliced (page 50) and twist the nut on as shown in Step 4.

JOINING SOLID TO STRANDED WIRE
To join stranded wire (often found on light fixtures and specialty switches) to solid-core wire, give the strands several twists between your thumb and forefinger to consolidate the strands. Then wrap the stranded wire around the solid wire, again with your fingers. Check that the stranded wire protrudes past the solid wire ⅛ inch or so. Twist on a wire nut, and tug both wires to make sure you have a solid connection. Finally, wrap the bottom of the wire nut with electrician's tape.

JOINING WIRE TO A TERMINAL

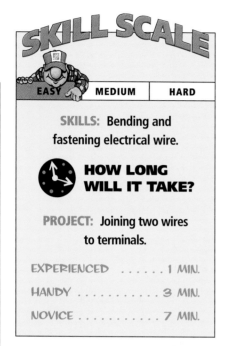

SKILL SCALE

EASY	MEDIUM	HARD

SKILLS: Bending and fastening electrical wire.

HOW LONG WILL IT TAKE?

PROJECT: Joining two wires to terminals.

EXPERIENCED ...... 1 MIN.

HANDY .... 3 MIN.

NOVICE .... 7 MIN.

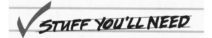

✓ STUFF YOU'LL NEED

TOOLS: Longnose pliers, side-cutting pliers, wire-bending screwdriver

MATERIALS: Wire, device with terminals

BASIC TOOLS AND SKILLS

Joining wire to a terminal is an important skill and a key step in most electrical projects. Do this step properly to ensure the device works and doesn't develop a short.

MAKING THE RIGHT CONNECTION

Electricians wrap the wire nearly all the way around the screw to make a connection that is completely reliable. With some practice, you can make joints just as strong. Bend a wire in a quarter circle, slip it under the screw head, and tighten the screw.

Many devices come with terminal screws unscrewed. Screw in any unused terminal screws so they won't stick out dangerously, creating a shock hazard should the terminal touch a metal box.

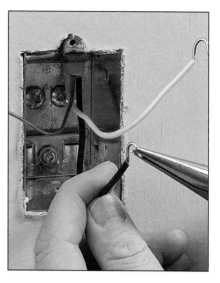

1 **START A LOOP.** Check that power is shut off. Strip about ¾ inch of insulation from a wire end (page 46). Using longnose pliers or the tip of a pair of combination strippers, grab the wire just above the insulation and bend it back at about a 45-degree angle. Move the pliers up about ¼ inch beyond the insulation, and bend again in the opposite direction, about 90 degrees.

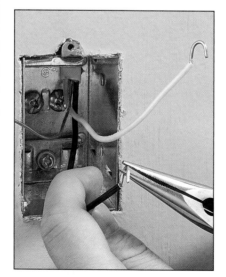

2 **OPTION A:**
BEND A QUESTION MARK. Use a longnose pliers to form a near-loop with an opening just wide enough to slip over the threads of a terminal screw. Move the pliers another ¼ inch away from the insulation, and bend again to form a shape that looks like a question mark.

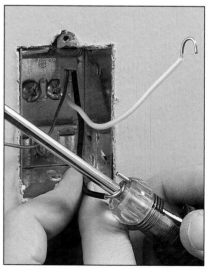

OPTION B:
USE A WIRE-BENDING SCREWDRIVER. This simple tool makes perfect hooks every time. Just push the stripped wire between the screwdriver shaft and the stud at the base of the handle. Twist the handle to make a perfect loop.

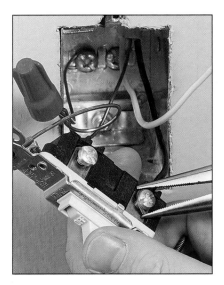

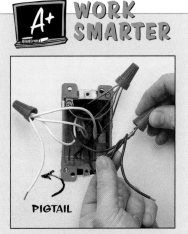
3 **SQUEEZE THE LOOP AROUND THE SCREW.** Make sure the terminal screw is unscrewed enough to become hard to turn. Slip the loop over the screw threads, with the loop running clockwise. Use longnose pliers or combination strippers to squeeze the loop around the terminal, then tighten the screw.

4 **WRAP WITH TAPE.** After all the wires are connected to a switch or receptacle, wrap electrician's tape around the body of the device to cover the screw heads and any exposed wires. The tape not only ensures that the wires stay attached, it keeps the terminals from touching the box.

PIGTAIL

USING PIGTAILS

Codes prohibit attaching two wires to one terminal. If you need to attach two wires to one terminal, cut a "pigtail" wire 6 inches long, and strip both ends. Splice the two wires to the pigtail, and join the pigtail to the terminal.

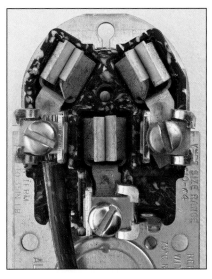

CONNECTING TO A 240-VOLT RECEPTACLE. Be certain that power is shut off—there is a dangerous level of power here. Strip about ½ inch of insulation from the wire end. The wire should be straight, not looped. Loosen the setscrew, poke the wire into the hole, and tighten the screw. (See page 145 for installing a 240-volt receptacle.)

SAFETY ALERT!

SKIP THE PUSH-IN OPTION

Most professionals don't trust this method even though it saves time. Many receptacles and switches have holes in the back for easy connection of wires. Once you've stripped the insulation (a strip gauge shows you how much),

you poke the wire in. To remove a wire, insert a small screwdriver into a nearby slot. The wire releases.

The system works, but the resulting electrical connection is not as secure as a connection made using a terminal screw. Take the extra minute to do it right.

WIRE NUTS AND TAPE

Wire nuts must cap all wire splices. These nuts come in several sizes, identified by color. On the package you will find a chart telling how many wires of a given size the nut can handle.

In older homes, you may find spliced wire ends wrapped with rubberized tape that is covered with cloth friction tape. Electricians often wrapped these well, so you may find them difficult to unwrap. (Slice with a utility knife before unwinding.)

Small, colored wire nuts are often included with light fixtures. If they are all plastic (with no metal threads inside) or if it is a challenge to get them to twist on because they are too small, use orange nuts instead. Use **yellow** connectors for splices as small as two #14s or as large as three #12s. **Orange** nuts handle combinations ranging from two #16 wires up to two #14s. Use **green** wire nuts for ground wires only. The hole in the top allows you to make an instant pigtail, with one wire poking out. **Red** wire nuts will grab splices as small as two #12s and as large as four #12s.

BUYER'S GUIDE

GET THE GOOD TAPE
The inexpensive tape often found in large bins at a home center will do the job, but many electricians prefer to use professional-quality tape. It's thicker and has better adhesive.

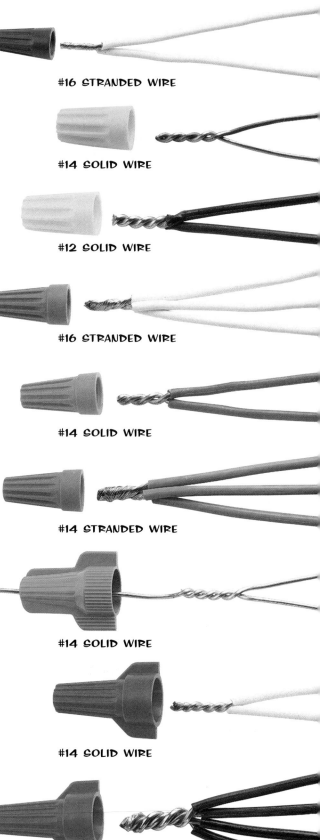

#16 STRANDED WIRE

#14 SOLID WIRE

#12 SOLID WIRE

#16 STRANDED WIRE

#14 SOLID WIRE

#14 STRANDED WIRE

#14 SOLID WIRE

#14 SOLID WIRE

#12 SOLID WIRE

BASIC TOOLS AND SKILLS

50

You don't have to be a professional electrician with a clipboard and a head full of electrical codes to correctly assess the safety of your home. Many problems that inspectors can find will be just as obvious to you once you learn how to find them.

This chapter begins with a simple "walk-around" inspection of your home, instructing you to examine fixtures, receptacles, and switches in plain sight. Later, the chapter demonstrates how to open your service panel and electrical boxes to inspect wires and terminals.

Each section identifies common household problems and refers you to a page describing how they can be fixed.

CHAPTER FOUR TOPICS

CHASE DOWN LIGHT FLICKERS
Did you ever sit down to read a magazine only to have your floor lamp flicker off and on? Besides being annoying, such flickering indicates a poor connection or a damaged switch, causing a dangerous electricity arc. Don't live with it. Check the bulb, switch, cord, and plug.

MEETING CODE

INSPECTING YOUR HOME

Local building codes and regulations are imposed to protect you and your family from shock and fire, and to make sure your wiring works reliably for decades.

WHEN YOU DON'T NEED CODE

Codes change over the years as new hazards are discovered and new products are introduced. It's possible that some of the wiring in your home fails to conform to current regulations. Usually, that's not a problem, as long as it conforms to the rules that existed when the wiring was installed. But any new work, even if it connects to old work, must meet code.

If you repair a fixture or replace one fixture with another without running new cable, there's no need to consult codes. Even if local regulations require you to get a permit for every fixture replacement, most inspectors will not want to be bothered with such small changes.

WHEN TO CONSULT CODES

If, during the course of an installation or inspection, you see wiring that's improperly connected and you don't know how to fix it, or if you see wiring that you do not understand, call in a pro or check with your local building inspection department.

Whenever you install new service in your home (make an installation that involves running new cable), you must get a permit and be sure to work according to code. (See page 112 for tips on working with inspectors.)

THE CEC AND LOCAL CODES

Codes vary from area to area. In fact, sometimes the regulations in neighboring cities can differ. However, all local codes are based on the *Canadian Electrical Code* (CEC). The CEC provides precise details about materials and installation—sometimes far more than you need to know about residential electrical installations. Copies are often available at your local library. The CEC is updated every few years to reflect changes in both products and installation techniques.

As you consult these books, remember that local codes prevail. Your local building inspection department will probably have brochures or leaflets that describe the most common electrical codes for residences.

BUYER'S GUIDE

HIRING A PRO

Most professional electricians are qualified, are honest, and charge fairly. Unfortunately, a few take advantage of homeowners' lack of knowledge and general fear of electricity. Too often, unscrupulous contractors target the elderly.

Word-of-mouth can be a great way to find a reliable contractor, but even sharp consumers may be unaware that they have overpaid or have paid for shoddy work. For a large job, get quotes from at least three contractors. Their bids should include a list of "specs"— everything that will be installed and how it will be installed.

Check that the contractor is licensed for your area and is covered by insurance. This way, if there is a fire or if a worker is injured on your property, you will not be held liable. If the work involves running new cable, the contractor—not the customer—should get a permit.

Read the section in this book about the installation you will pay for. Don't be afraid to ask the electrician to explain what he or she is doing. Question everything that looks substandard. In particular, have the contractor explain how the installation is grounded.

A WALK-AROUND INSPECTION

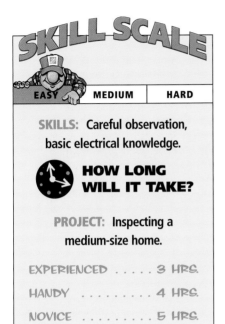

SKILL SCALE

| EASY | MEDIUM | HARD |

SKILLS: Careful observation, basic electrical knowledge.

HOW LONG WILL IT TAKE?

PROJECT: Inspecting a medium-size home.

EXPERIENCED ..... 3 HRS.

HANDY .... 4 HRS.

NOVICE .... 5 HRS.

STUFF YOU'LL NEED

TOOLS: Flashlight, receptacle analyzer

MATERIALS: Paper and pencil

Once you know what to look for, most household wiring problems are easy to find and fix. Make a whole-house inspection by checking closets, the attic, the basement or crawl space, and the garage. Globes for light fixtures will be the only things you have to remove. Prioritize everything that needs to be done and immediately take care of all potentially hazardous problems.

INSPECTING RECEPTACLES FOR PROBLEMS

1 **LOOK FOR DEVICES THAT LACK COVER PLATES.** If the cover plate to a switch or receptacle is missing, replace it immediately. Lack of cover plates presents a dangerous situation: Children might reach into the electrical box, where live wires lurk. An adult fumbling for the switch at night could receive a shock as well. Replace any missing cover plates.

CRACK

2 **BEWARE OF CRACKED RECEPTACLES.** A crack on the outside may mean that a receptacle's inner circuitry is in danger of shorting out. New receptacles are inexpensive and easy to replace (page 26).

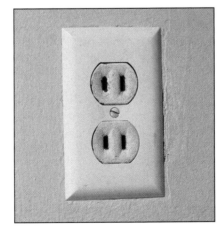

3 **UPGRADE UNGROUNDED RECEPTACLES.** Receptacles with only two slots and no grounding hole are ungrounded. Homeowners got by with ungrounded wiring for decades, but grounding provides a necessary level of protection. If you can't install grounded receptacles, add ground-fault circuit interrupter (GFCI) receptacles (page 27).

SAFETY ALERT!

NOT ON SAFE GROUND!
Attaching the adapter to the screw that secures the cover plate doesn't guarantee that the receptacle is grounded! **If the receptacle itself isn't grounded through the system, screwing in the adapter will only give you a false sense of security and in many locations is actually illegal. See page 13 for information on proper grounding procedures.**

INSPECTING YOUR HOME

5 **WIGGLE SWITCH TOGGLES.** If there seems to be too much play in the switch toggle—especially if you hear a pop when the switch is turned on or off—the device should be replaced (pages 22–24).

6 **TEST A GFCI.** Just because a ground-fault circuit interrupter (GFCI) receptacle is supplying power doesn't mean it will protect against shock. A GFCI can lose its protecting capacity. Test each GFCI by pushing the test button. The reset button should pop out. If it doesn't, replace the GFCI receptacle (page 27).

7 **STANDARD RECEPTACLES IN DAMP AREAS.** A wet receptacle is a shock hazard, so current codes call for ground-fault circuit interrupters (GFCIs) in bathrooms, near sinks, and outdoors. See page 27 for instructions on installing a GFCI. (Canada: GFCIs are not allowed on kitchen counters. Use split receptacles instead. See pages 144, 175.)

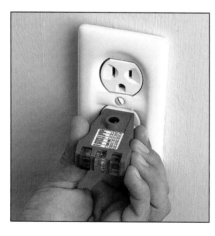

8 **CHECK FOR GROUNDING AND POLARIZATION.** If a receptacle analyzer indicates that a receptacle is not grounded, shut off the power and remove the cover plate and the receptacle (page 26). Compare the wiring with the examples shown on pages 11 and 13. If a wire is loose, reattach it. If a receptacle is not polarized, switch wires so that the hot wire is connected to the brass terminal and the neutral wire is connected to the silver terminal. If you are not sure what is wrong, call a pro.

SAFETY ALERT!

PLASTIC INSERT

COVER ROTATES

KID-SAFE RECEPTACLES
Although kids sometimes pull them out, the simplest and cheapest protection is to push a plastic insert (above left) into each unused outlet. For more reliable protection, install a special cover plate (above right) that must be twisted before inserting the plug. Most importantly, teach children to respect electricity and to stay away from all receptacles.

INSPECTING FIXTURES AND BOXES FOR PROBLEMS

CHECK YOUR INSURANCE
Some insurance policies have exclusions stating that the insurer does not have to pay for fire damage if the home had certain defects in its wiring. It's in your financial interest, as well as in the interest of safety, to make sure your electrical system is free of obvious defects.

1 **CHECK FOR UNSECURED GLOBES.** After replacing a light bulb, it's easy to tighten a setscrew before the globe is properly nested into place. With a little vibration, the globe could crash to the floor. When replacing a globe, unscrew the setscrews a bit more than necessary for removing the globe. Slip the globe up and make sure its lip is above all the screws before you tighten them. Check again after tightening.

2 **DETERMINE IF THE BULB WATTAGE MATCHES THE FIXTURE.** It's easy to overlook the stickers inside light fixtures that state the maximum allowable wattage. Bulbs with too-high wattage will overheat fixtures. At best, you'll have to change bulbs more often; at worst, overheating can cause a fire. If you need more light, install a new fixture (pages 28–29) with a higher wattage allowance.

If you find a problem, be sure to shut off the power before fixing it.

3 **AVOID BARE LIGHT BULBS IN CLOSETS.** Too often, light fixtures in closets don't have globes. Sweaters, comforters, cardboard boxes, and other flammables placed too near bare bulbs can catch fire. The best solution is to replace your closet light with a fixture that has a globe covering the bulb. Or install a surface-mounted fluorescent light (page 104).

4 **MAKE SPACE IN CROWDED BOXES.** If a junction box is so crowded that it prevents the cover plate from being tightened all the way, install a box extender (page 82), or replace the box with a larger one.

CHECKING CORDS AND WIRES FOR PROBLEMS

1 **CHECK FOR BROKEN OR BENT GROUNDING PRONGS.** Appliance and tool grounding plugs are installed for your safety. Do not remove or bend back grounding prongs—you will negate an important safety feature. (See page 11 for an explanation of grounding.) Replace a plug that has a bad prong (pages 72–73).

Don't just tape a damaged wire — replace it! And, never run an extension cord under a carpet or area rug!

2 **DON'T OVERLOAD RECEPTACLES.** This many-armed monster is awkward and unsafe. Using too many appliances at once can overheat the receptacle. Install another receptacle (page 141).

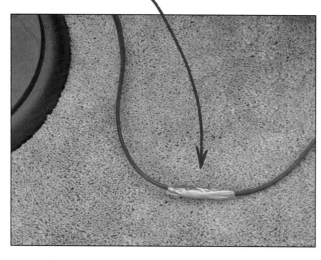

3 **WATCH OUT FOR DAMAGED CORDS.** A cord with less-than-perfect insulation can cause shock or start a fire. All lamp cords and appliance cords should be free of nicks; you should see no bare wire. Run your fingers along each unplugged cord. If you feel cracks or if the cord is brittle, replace it (pages 68–69). Pay special attention to the cord near the plug, where insulation is most often damaged.

BUYER'S GUIDE

TAME THOSE CORDS
A tangle of cords near a desk can become unwieldy, and a stray cord poses a tripping danger. Run the cords through a plastic sleeve or corral them with a removable strap.

INSPECTING YOUR HOME

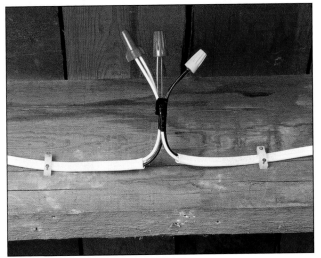

4 **CHECK CABLE ENTERING A BOX WITHOUT A CLAMP.** Cable and wire must be firmly held, because vibration can cause rubbing—which can harm insulation. Metal boxes in particular have sharp edges that can nick insulation. (Plastic boxes do not usually require clamps. Staple the cable to a stud or joist within 12 inches of the box.) Shut off power to the box, unhook the wires, and attach the cable with a cable clamp (pages 123 and 125).

5 **FIX EXPOSED SPLICES.** Exposed connections can easily be bumped and loosened, running the risk of a short or fire. That's one of the reasons all wire and cable splices must be within an approved electrical box—either a junction box, a switch or receptacle box, or a fixture that is designed to be used as an electrical box. (To add a box, see pages 118–119 and 137.)

6 **SECURE LOOSE CABLE. NEVER USE CABLE AS A HANGING ROD.** Codes in some areas permit exposed NM (nonmetallic) cable in basements and garages, while other areas require armored cable or metal conduit. Whatever type of cable you have, it should be tightly stapled to a surface so it cannot accidentally be pulled out.

7 **CHECK KNOB-AND-TUBE WIRING.** This old style of wiring is still in use in many homes. As long as the wires are completely undisturbed and the wire insulation is in good shape, it can be used. But the insulation can get brittle and easily damaged. Have a pro evaluate it for safety. If you ever replace or extend this type of wiring, do not use more knob-and-tube hardware. Instead, use standard cable clamped to electrical boxes (pages 122–127).

INSPECTING BOXES FOR PROBLEMS

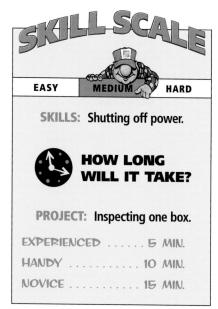

SKILLS: Shutting off power.

HOW LONG WILL IT TAKE?

PROJECT: Inspecting one box.

EXPERIENCED ...... 5 MIN.

HANDY .... 10 MIN.

NOVICE .... 15 MIN.

B e cautious when opening and inspecting a box for potential problems. Kill power to the box at the service panel before you begin work. Remember, however, that there is always the potential that more than one circuit goes to a box so work carefully. Use rubber-gripped tools, wear rubber-soled shoes, and do not touch bare wires. See pages 22 and 26 for instructions on opening a switch box and a receptacle box.

STUFF YOU'LL NEED

TOOLS: Voltage tester or multitester, screwdriver, flashlight

MATERIALS: None required

LOOK FOR OVERCROWDING. Using a rubber-handled screwdriver, back out the screws holding the box cover plate until the cover is loose enough to remove. Too many wires crammed into too small a box can lead to shorts. Also note that the box is not grounded and should be replaced. See pages 118–119 to select the right size box for the job.

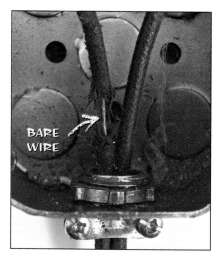

CHECK FOR OLD, CRACKED INSULATION. If wire insulation is hard and brittle, shut off the circuit and wrap the damaged insulation with a hot-shrink sleeve (page 82). If all the wiring in your house has brittle insulation, you may need to hire an electrician to rewire your house.

WORK SMARTER

OPENING A JUNCTION BOX

Junction boxes have flat metal cover plates and are usually found in basements, garages, or utility rooms. They generally hold six or more wires spliced with wire nuts. If possible, trace the cables from the junction box back to the service panel. Follow the hot wires in the service panel to figure out which circuit or circuits need to be shut off.

It may not be possible for you to shut off the power before opening a junction box, if, for example, you can't follow the cable. Also, wires from two or more circuits may run through a single junction box. So, even if you've shut off power, act as if power is still on. Loosen the two screws holding the cover plate, and ease off the plate. If you need to test for power, gently pull out wires so that no two splices are closer together than 1 inch. Unscrew the wire nuts and touch the probes of a multitester or voltage tester to both neutral (white) and hot (black or colored) wires (pages 44–45).

LOOK FOR EXPOSED WALL MATERIAL AROUND BOXES. An electrical box should be flush with the finished wall; if not, it poses a fire hazard. Also note that the box is not grounded and should be replaced. To solve this problem install a grounded box that fits flush to the surface.

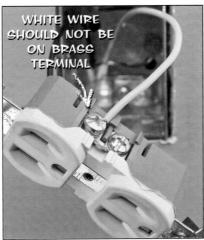

CONFIRM POLARIZATION—WHITE WIRES GO TO SILVER, BLACK TO BRASS. If the white wire is connected to a brass terminal and the black one is connected to a silver terminal, the receptacle isn't polarized. An appliance or light plugged into it may be energized even when switched off. Reverse the wires so white goes to silver and black goes to brass. (See page 11.)

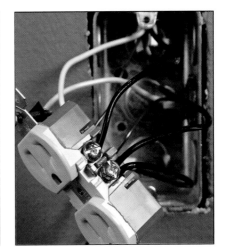

ENSURE SAFETY WITH A PIGTAIL CONNECTION. Two wires should not be attached to the same terminal: Not only do they make a poor connection, they can pop off and short. Remove the two wires and make a pigtail connection (page 49). Also note that the box is not grounded and should be replaced.

CHECK ARMORED CABLE CONNECTORS. The cut ends of armored cable are sharp and can slice through wire insulation. Even if no damage has been done, install a plastic bushing (page 125) wherever one is missing. If a wire has been nicked, cover it with a hot-shrink sleeve (page 82).

ENSURE THAT BOXES ARE SECURE. When electrical boxes are not securely anchored, wiring or connections can be damaged. If a loose switch or receptacle box is next to a stud, pull out the device, drill a hole through the side of the box, and drive a screw through the box and into the stud. If it is not near a stud, replace the box with an "old-work" box that clamps to the drywall or plaster (pages 133–134).

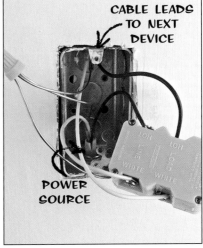

CHECK THAT GFCI RECEPTACLES ARE CORRECTLY WIRED. The wires coming from the power source should connect to the LINE terminals, and the wires leading out to other receptacles or fixtures should connect to the LOAD terminals (page 27). Also note that the box is not grounded and should be replaced.

INSPECTING A SERVICE PANEL

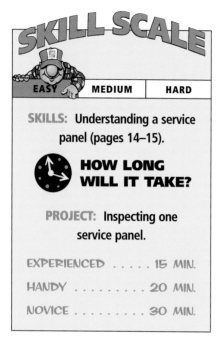

SKILL SCALE

| EASY | MEDIUM | HARD |

SKILLS: Understanding a service panel (pages 14–15).

HOW LONG WILL IT TAKE?

PROJECT: Inspecting one service panel.

EXPERIENCED . . . . . 15 MIN.

HANDY . . . . 20 MIN.

NOVICE . . . . 30 MIN.

✓ STUFF YOU'LL NEED

TOOLS: Screwdriver, flashlight, voltage detector

MATERIALS: None required

Even if your service panel was installed correctly, substandard wiring may have been added later. Inspect the entire panel, but pay special attention to new additions. Look for a melted breaker, burned wires, or a burned bus bar. Call an electrician if you see any sign of scorching or overheating. If you can't read the size of an old wire, carefully compare the thickness of its copper with newer wires. If you see three or more wires attached to a breaker, call in an electrician: Codes might allow you to make a pigtail connection (page 49) inside a panel.

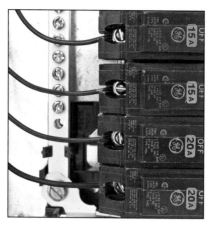

CHECK WIRE THICKNESS. A #14 wire connected to a 20-amp breaker poses a very dangerous situation. A 20-amp breaker is designed for a #12 wire or larger. A #14 wire can overheat and even melt insulation or start a fire before the 20-amp breaker trips. Replace the breaker with one that is 15 amps. Consult a professional electrician if this causes the breaker to trip often.

SAFETY ALERT!

OPENING A PANEL

Study the safety precautions on page 15 before opening a panel. Switch off the main power in the box before attempting any work. **The outer cover includes the door. Loosen or remove screws at the bottom, sides, and top. Lift out the cover.**

You'll probably see the wires and their connections to the breakers and the neutral bar. Remove the second cover if you need to remove a breaker. Don't touch any wires.

TROUBLESHOOTING YOUR FUSE BOX

Check your service panel for these signs of trouble:

■ **Rust** in the fuse or breaker box may indicate that the box is getting wet. This can be very dangerous. Make sure the box is dry at all times.

■ A **30-amp fuse** may indicate that a higher-than-recommended fuse was installed because that circuit kept blowing fuses. Also, check 20-amp fuses to make sure they shouldn't be 15-amps. If the wire leading to the fuse is not #10 or thicker, the fuse should be lower in amperage—15 amps for #14 wire and 20 amps for #12 wire.

■ Constantly **blown fuses** are an annoyance—and they indicate that a circuit is overstressed. See pages 62–63 for tips on balancing circuit loads.

■ Without an **index**, or circuit map, it can be challenging to figure out which circuit to shut off or turn back on. See opposite page for how to map circuits.

■ In some areas, a panel attached to a **flammable surface** is considered a fire hazard; nonflammable material is required between the panel and wood.

■ **Open knockouts** also present a fire danger. Buy push-in "goof plugs" designed to fill open knockouts.

MAPPING CIRCUITS

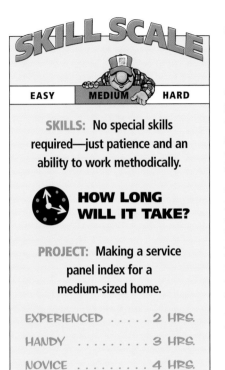

SKILL SCALE

EASY | MEDIUM | HARD

SKILLS: No special skills required—just patience and an ability to work methodically.

HOW LONG WILL IT TAKE?

PROJECT: Making a service panel index for a medium-sized home.

EXPERIENCED . . . . . 2 HRS.

HANDY . . . . 3 HRS.

NOVICE . . . . 4 HRS.

✓ STUFF YOU'LL NEED

TOOLS: 2 cordless phones or walkie-talkies, circuit finder

MATERIALS: Masking tape, marker, pencil, and paper

When making a repair or new installation, knowing which circuit controls which outlet speeds the job and makes working safer. That's why electrical codes require service panels to have an index telling which receptacles, lights, and appliances are on which circuit. If your panel has no index, creating one will take some time. Prepare by turning on all the lights in the house. Plug a light, fan, or radio into as many receptacles as possible, and switch them on. Turn on the dishwasher and open the door of the microwave oven. With the whole house switched on, you are ready to map.

1 MAP THE LOCATIONS OF ALL OUTLETS. Draw a rough sketch of each floor in your house, noting the location of every receptacle, switch, light, and appliance. (You may want to use the symbols shown on page 114.) On the service panel, place a numbered piece of tape next to each breaker or switch.

2 IDENTIFY CIRCUITS. To communicate with your helper, use a pair of walkie-talkies or two cellular phones. Start at the top of the panel. Switch off the circuit and have your helper identify the room without power. On the map, jot down the number of the circuit next to each outlet that is turned off. Repeat these steps for each circuit.

3 DRAW UP THE INDEX. Write an index that accounts for all the fixtures, receptacles, and hardwired appliances in your home. Attach the index to the inside door panel. You may be surprised to find that some circuits wander through several rooms. This can be confusing, but it is not dangerous.

AVOIDING CIRCUIT OVERLOADS

INSPECTING YOUR HOME

The total power used by all of a home's light fixtures, lamps, appliances, and tools is called "demand." When demand exceeds the safe capacity of a circuit, the circuit is overloaded.

BREAKERS AND FUSES

It's usually easy to tell if a circuit is overloaded: The breaker trips frequently or the fuse keeps blowing. This probably means that wires are overheating, posing a threat to your home.

Sometimes the solution is simple: Move one high-amperage appliance (such as a microwave oven or toaster) to a receptacle on another circuit. If the overloads stop, then the problem is solved. If not, you may need to install a new circuit (pages 168–169).

Overloading problems often occur on 120-volt circuits, which serve multiple receptacles and lights. Most 240-volt circuits serve only one receptacle or appliance.

If a 240-volt circuit regularly overloads, change the wiring.

To better understand troublesome circuits and to prepare for adding new electrical lines, the chart below shows how close the circuits are to being overloaded.

CHECKING WATTS AND AMPS

If the service panel does not have an accurate index, map the house and add an index (page 61). Find a circuit's amperage rating by looking at the circuit breaker or fuse. Add up the wattage of every light bulb on the circuit. Note the amperage or watt rating for every appliance and tool plugged into receptacles as well. This information should be printed somewhere on the appliance. Examples are illustrated opposite. Some appliances vary widely in ratings, so check appliances individually. Older appliances usually have a higher rating.

SAFE CAPACITY

Codes require that appliances and fixtures on a circuit do not exceed "safe capacity," usually defined as the total capacity minus 20 percent. (See the chart, below left.) If the total demand exceeds a circuit's safe capacity and you can't solve the problem by plugging an appliance into a receptacle on another circuit, install a new circuit (pages 168–169). If a circuit suddenly becomes touchy—tripping the breaker at the slightest provocation—check to see if the breaker is functioning correctly (page 81).

CALCULATING CIRCUIT CAPACITY

Here are two ways to calculate circuit capacity. First, if you know the amperage and voltage of the circuit, you can determine the total capacity by doing this calculation:

Amps × Volts = Watts

For example, if you have a 15-amp, 120-volt circuit, total capacity in watts is 15 × 120, or 1800 watts (15 × 120 = 1800). With a 20-amp, 120-volt circuit, total capacity is 2400 watts (20 × 120).

Or work the other way around:

Watts ÷ Volts = Amps

If all the bulbs in a pendant light fixture add up to 600 watts, the light is using 5 amps (600 ÷ 120 = 5). If such a fixture hangs in your kitchen, don't run a toaster (at 6–13 amps) on the same 15-amp circuit or you might overload the circuit.

Some electricians use this general rule: Allow 100 watts for each amp. That means allowing no more than 1500 watts on a 15-amp circuit and no more than 2000 watts on a 20-amp circuit.

WORK SMARTER

SAFE CAPACITY FOR 120-VOLT CIRCUITS

To be sure your circuit won't overload, check individual appliances to determine the watts required by each appliance and fixture on a circuit. Total the usage to make sure it is within the safe capacity shown here.

AMPS	TOTAL CAPACITY	SAFE CAPACITY
15A	1800 watts	1440 watts/12 amps
20A	2400 watts	1920 watts/16 amps
25A	3000 watts	2400 watts/20 amps
30A	3600 watts	2880 watts/24 amps

WATTAGE AND AMPERAGE RATINGS

These ratings are examples only. Check appliances individually.

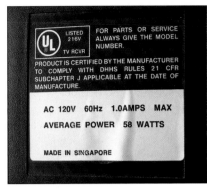

TELEVISION
50–300 watts/ 0.4–2.5 amps

CIRCULAR SAW
1200 watts/ 10 amps

ALLOW FOR MOTOR SURGE

During the first few seconds a motor is started, it uses significantly more power than during normal operation. A circuit supplying appliances with motors—a refrigerator, freezer, or air-conditioner, for instance— needs extra capacity to handle occasional power surges.

REFRIGERATOR
700–1200 watts/ 5.8–10 amps

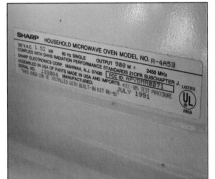

MICROWAVE OVEN
900–1500 watts/ 7.5–12.5 amps

Homer's Hindsight

WATT CHEER!

I covered the house with Christmas lights last year, and without realizing it, really created an overload. I plugged in the extension cord for the lights and the whole circuit blew. I added up the demand and found that my 10 strings of lights pulled 1800 watts. And I was about to hook them to a 600-watt timer! I switched to another circuit. This summer, I'm installing an outdoor circuit just for decorative lights.

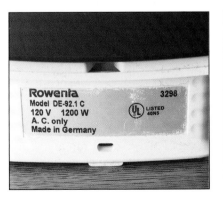

IRON
1000–1200 watts/ 8.3–10 amps

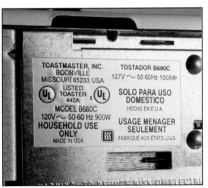

TOASTER
800–1600 watts/ 6.6–13.3 amps

INSPECTING YOUR HOME

ALUMINUM WIRING

When copper prices increased in the early 1970s, builders in many areas switched to aluminum wire. Homeowners soon discovered, however, that aluminum posed a fire hazard, especially when connected to brass or copper terminals or wires. By the time aluminum wire was banned, thousands of homes had been wired. Because aluminum expands and contracts over time, it can loosen from terminals, causing faults. Also, where aluminum is attached to brass or copper, it oxidizes, degrading the connection.

Consider replacing aluminum wires with copper if the wires run through conduit or Greenfield. You (or an electrician) can install new wires by attaching them to the old wires and pulling them through the conduit (page 127). However, if the aluminum is encased in NM or armored cable, replacing it will be difficult and costly.

PREVENTIVE MAINTENANCE FOR ALUMINUM WIRING

To check the condition of your aluminum wiring, you'll need to systematically open every switch box, receptacle box, fixture, hardwired appliance, and junction box in your home.

Shut off power to a circuit and open its boxes. If a switch or receptacle has CO/ALR written on it (center right), it is safe to connect aluminum wires to it. If the device is a standard receptacle, replace it with a CO/ALR device. (Buy them at electrical supply stores if your home centers do not carry them.) How to replace devices is described on pages 22–23 and 26.

Or, use a more time-consuming but less costly method (bottom right): Disconnect an aluminum wire from its terminal, and connect it to a short pigtail made of copper. Squirt antioxidant from its tubular container (below right) onto the wire ends. Then, twist the wires together and attach an Al/Cu wire nut, which is made for this purpose. Connect the copper pigtail to the standard switch or receptacle.

Make sure that throughout the house all aluminum-to-copper or aluminum-to-brass connections are brushed clean of corrosion (look for a powdery white coating, especially on devices near damp areas); coat the connections with antioxidant. Check each device, backing off the terminal screws, adding the antioxidant, and firmly retightening the screws. Aluminum breaks easily. If a wire end is cracked, snip it off and restrip.

Check all connections to terminals annually, and tighten them as needed.

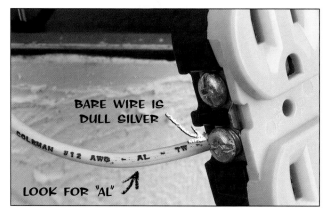

CHECK WIRING: Aluminum wire is marked "AL." The stripped wire is a dull silver color. Aluminum is a soft metal and strains easily. Look for cracks in the bare wire.

CHECK DEVICES. All switches and receptacles should have CO/ALR stamped on them. If not, replace them with a CO/ALR device. Check every switch and receptacle.

APPLY ANTIOXIDANT. To keep the aluminum from developing a nonconductive layer of oxidant, especially where aluminum is joined to copper, snip off bare wire ends, restrip, and coat the wires with antioxidant. Use Al/Cu wire nuts, and attach to copper pigtails.

5 ELECTRICAL REPAIRS

When electrical devices and fixtures no longer work, often the logical solution is to replace rather than repair them. Switches, receptacles, lamps, and overhead lights may not cost enough to warrant the time it takes to diagnose and repair them.

Some repairs, however, are quick and easy. You may be able to get your lamp to work again by simply pulling up the tab on the light socket (see page 66). If you have a valuable antique lamp or overhead light—a treasured part of your home—you certainly have a vested interest in getting it back into working order.

ELECTRICAL REPAIRS

CHAPTER FIVE TOPICS

FIXING LAMP SOCKETS

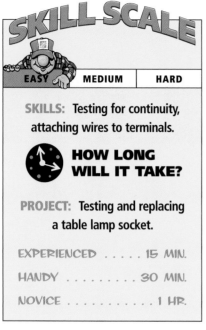

ELECTRICAL REPAIRS

✓ STUFF YOU'LL NEED

TOOLS: Screwdriver and a continuity tester or multitester

MATERIALS: New socket if needed, and electrician's tape

If a lamp doesn't work, eliminate the obvious causes first. Make sure the lamp is plugged in. Make sure the bulb is OK. A burned-out bulb usually makes a tinkling sound when you shake it. Screw in a fresh bulb if necessary.

Check that the receptacle's circuit hasn't blown a fuse or popped a breaker. If the lamp still doesn't work, test by plugging in a lamp that you know is in working order. If it lights up, you've isolated the problem to the lamp itself. Before replacing the cord or switch, take a look at the socket.

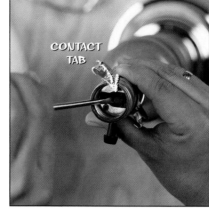

1 PRY UP THE CONTACT TAB.
Unplug the lamp and remove the bulb. If the contact tab is corroded or rusty, scrape it with a screwdriver. If the tab lies flat, it may not be making solid contact with the base of the bulb. Gently pry up the tab about ⅛ inch and retest. If the lamp still doesn't work, go to the next step.

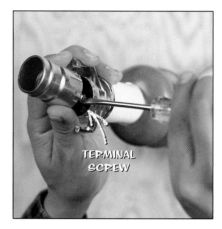

2 REMOVE THE SOCKET. Look for the word PRESS on the socket shell. Push there with your thumb as you squeeze the shell and wiggle it up and out. If there is a cardboard sleeve, remove it, too. Loosen the two terminal screws and pull out the socket.

Connect the ridged (neutral) wire to the silver terminal and the smooth (hot) wire to the brass terminal.

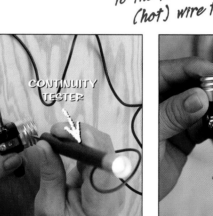

3 TEST THE SOCKET AND SWITCH. Test the socket with a continuity tester (shown) or a multitester (pages 44–45). Touch one probe to the neutral (silver) screw and the other to the threaded metal of the socket. If the tester bulb doesn't light, replace the socket. If the socket has a switch, touch the clips to the brass terminal and to the contact tab. If the switch is defective, replace it. If it is not, test the cord and plug (page 68).

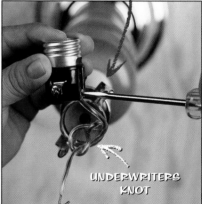

4 REPLACE THE SOCKET. You may need to loosen a small setscrew in order to unscrew the old socket base. Install the new base, threading the cord carefully so you don't nick the insulation. Tie the wires with an Underwriters knot as shown. Twist the strands together with your fingers, and form a partial loop. Wrap each wire clockwise around a terminal, and tighten its terminal screw. Slip on the cardboard sleeve, and snap down the socket shell into position.

REPLACING LAMP SWITCHES

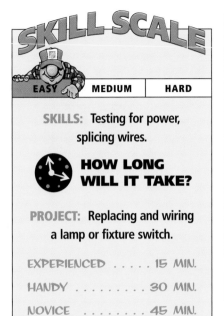

SKILL SCALE

EASY	MEDIUM	HARD

SKILLS: Testing for power, splicing wires.

HOW LONG WILL IT TAKE?

PROJECT: Replacing and wiring a lamp or fixture switch.

EXPERIENCED . . . . . 15 MIN.

HANDY . . . . 30 MIN.

NOVICE . . . . 45 MIN.

✓ STUFF YOU'LL NEED

TOOLS: Voltage tester, combination strippers, pliers

MATERIALS: New switch, electrician's tape, wire nuts

A toggle, pull-chain, or twist switch is not an integral part of the lamp or the fixture on which it's mounted. It's an inexpensive switch that can be easily replaced—and may need to be replaced yearly, if heavily used.

There's one universal hole size, so you can interchange twist switches with toggles or pull-chains.

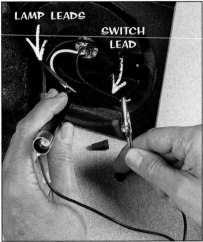

LAMP LEADS   SWITCH LEAD

1 **TEST THE SWITCH.** Unplug the lamp. Remove the bottom of the lamp to access the wiring. Remove one of the wire nuts that connects a lead from the switch to the lamp wires. With the bulb still in the socket, clip one probe of a continuity tester to the switch lead and the other to the lamp wires. Try the switch several times. If the continuity detector doesn't light, the switch is defective.

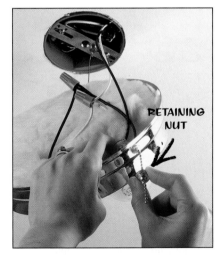

RETAINING NUT

REPLACING A PULL-CHAIN SWITCH.
Test a switch like this (common on ceiling lights and fans) in the same way as a toggle or twist switch. It mounts with a retaining nut. Some porcelain ceiling fixtures have built-in switches; these can't be repaired.

RETAINING NUT

2 **REPLACE THE SWITCH.** Unplug the lamp. Unscrew the switch retaining nut—you may need to use pliers. Unravel the wires, then pull out the switch. If the wires on the lamp are damaged, snip off the stripped portion and restrip the insulation. Insert the switch into the hole, and tightly screw in the retaining nut. Splice the switch leads to the lamp, and twist on the wire nuts.

A+ WORK SMARTER

WIRING A THREE-LEVEL SWITCH

If a fixture-mounted switch powers a light or fan at more than one level, the wiring is more complicated. If the switch has more than two leads, carefully tag the lamp or fixture wires with pieces of marked tape so you know which wire goes where when you install the replacement. Take the old switch with you to the hardware store or home center to buy an exact replacement.

REWIRING LAMPS

ELECTRICAL REPAIRS

SKILL SCALE

EASY	MEDIUM	HARD

SKILLS: Stripping wire, testing for continuity.

HOW LONG WILL IT TAKE?

PROJECT: Rewiring a lamp.

EXPERIENCED . . . . . 15 MIN.

HANDY . . . . 30 MIN.

NOVICE . . . . 1 HR.

✓ STUFF YOU'LL NEED

TOOLS: Multitester or continuity tester, combination strippers, pliers, screwdriver, utility knife

MATERIALS: New lamp cord, electrician's tape, wire nuts

Wiring lamps is simple work. Electricity travels up through the lamp body through a cord until it reaches the socket. If the tests show that the socket works okay (page 66), the problem is probably with the cord. Don't repair a section of a cord: Cord splices never look good and they unravel easily. Install a new replacement lamp cord, which has a molded plug.

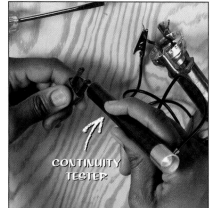

CONTINUITY TESTER

1 **TEST WIRES FOR CONTINUITY.** With the socket removed (page 66), touch the probes of a continuity tester or multitester (pages 44–45) to the end of the ridged (neutral) wire and the wide prong of the plug. Then touch the probes to the smooth (hot) wire and the narrow prong. If either test fails to show continuity, replace the cord and plug. If the prongs are the same size, test each wire with both prongs. The meter should show continuity on one prong only.

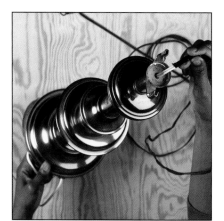

3 **PULL THE NEW CORD THROUGH.** This step is easier with a helper. While pulling up on the old cord at the top of the lamp, feed the new cord into the hole at the base. If the tape gets stuck, pull the cord out and wrap the tape more tightly. Keep pulling until the new cord emerges from the top. Unwrap the tape. Tie an Underwriters knot, and connect the new cord to the socket (page 66).

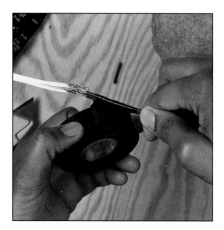

2 **TO PULL THE NEW CORD THROUGH THE LAMP, HOOK THE NEW CORD TO THE OLD.** Cut the old cord about 8 inches past the lamp base. Strip the ends of the old and new cords, and twist all the strands clockwise with your fingers so that no strands are loose. Bend the old wires and the new wires. Hook them together as shown. Wrap the joint tightly with electrician's tape.

SAFETY ALERT!

EXTRACTING A BROKEN BULB

If a bulb is broken and stuck in the socket, don't try to unscrew it by hand. Unplug the lamp. Press a potato onto the broken glass and then twist. Or, insert the end of a wooden broom handle into the middle of the socket and twist.

REPAIRING A TWO-SOCKET LAMP

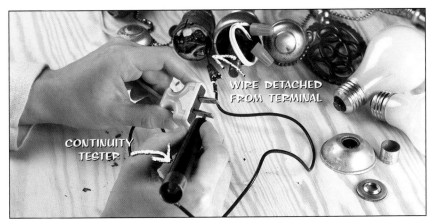

WIRE DETACHED FROM TERMINAL

CONTINUITY TESTER

TWO LEADS OF ONE CORD

1 **REMOVE THE COVER AND TEST THE LAMP.** If a lamp has two or more sockets and only one doesn't operate, test and replace the defective one as you would a one-socket lamp (page 66). Remove the cover plate and make sure the wire connections are tight. With the lamp switch on, use a multitester or continuity tester to check for continuity (pages 44–45). If only one socket fails to light, the wire between the splice and the socket is probably the culprit. If all the sockets fail to work, then the cord between the plug and the splice is bad.

2 **REWIRE THE LAMP.** Replace one cord at a time. For the sockets, cut and strip pieces of cord to the length of the old pieces. Connect the ridged (neutral) wire to the silver socket terminal and the smooth (hot) wire to the brass terminal. When splicing, always connect ridged wire to ridged and smooth to smooth.

DIVIDING AND STRIPPING LAMP CORD

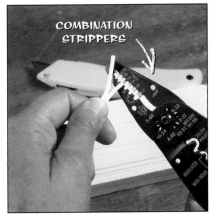

COMBINATION STRIPPERS

1 **DIVIDING THE LAMP CORD.** Separate part of a lamp cord into two wires before making connections. Stick the tip of a knife blade into the little valley between the two cords, and push down until it jabs firmly into the work surface below.

2 **STRIPPING THE LAMP CORD.** Pull the cord—not the knife—to separate the wires. Once you have made this cut, pull the wires farther apart if needed. Use combination strippers to remove insulation. Work carefully so that you don't pull off more than a couple of wire strands with the insulation.

TRIP SAVER

A LAMP REWIRE KIT

Some lamps have special components—such as washers or plastic stoppers. Replace them while you are rewiring. A lamp rewire kit contains the cord with plug and the little parts unique to that kind of lamp. The kit shown is for a bottle-type lamp.

REWIRING DESK LAMPS

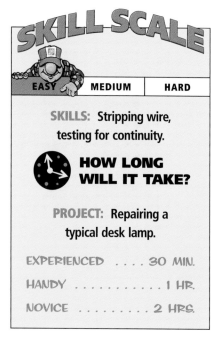

SKILL SCALE

| EASY | MEDIUM | HARD |

SKILLS: Stripping wire, testing for continuity.

HOW LONG WILL IT TAKE?

PROJECT: Repairing a typical desk lamp.

EXPERIENCED . . . . 30 MIN.

HANDY . . . . 1 HR.

NOVICE . . . . 2 HRS.

STUFF YOU'LL NEED

TOOLS: Voltage tester, continuity tester, combination strippers, pliers, screwdriver

MATERIALS: Electrician's tape, replacement lamp wire

R epairing incandescent desk lamps is similar to floor and table lamp repair. If a lamp goes out, first check that the bulb is good, that the cord is plugged in, and that the receptacle is getting power. (See page 78 to repair fluorescents.)

If a lamp has a fixture-mounted switch, check that it is working (page 67). Gently pry up the tab inside the socket (page 66) and try the lamp again. Test the cord for continuity (pages 44–45). Replace it if it fails the test or appears damaged. Then test the socket.

If you have an old fluorescent lamp, you probably won't find any replacement parts other than new bulbs. Parts vary according to the brand and model of the lamp. New sockets, starters, and ballasts simply aren't widely available.

1 **DISCONNECT THE SOCKET ON AN INCANDESCENT LAMP.**
Unplug the lamp. If there is a twist switch at the top of the shade, unscrew its retaining nut (see inset) and gently pull out the socket. If possible, push or pull wire for some slack to make it easier to get at the terminal screws. On some types of lamps, the socket can come out. Test the socket as shown on page 66.

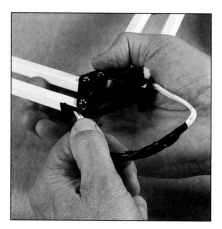

2 **REWIRE.** See pages 68–69 for general rewiring instructions. In the case of a desk lamp, there may be extra twists and turns. Feed the cord carefully through each one. Some models allow you to temporarily loosen joints to make it easier to pull the wire through. You may want to use a small amount of pulling lubricant (page 127).

CLOSER LOOK

REPAIRING HALOGENS
Halogen bulbs get extremely hot, so allow each to cool before removing it. Use a cloth—oil from your fingers can damage the bulbs. (If you touch the bulb, clean it with alcohol.) With the lamp plugged in and turned on,

test the socket. Set a multitester to test low voltage, and test the socket terminals where the bulb prongs connect. If you get the correct voltage, replace the bulb. If the reading is low, see if the lamp fuse is blown. If the these steps don't fix it, the transformer or the wiring is faulty; buy a new lamp.

REPAIRING PENDANT FIXTURES

Regular flush-mounted ceiling fixtures rarely need repair, and when they do, the wiring is straightforward. Pendent fixtures or chandeliers, however, often have a tangle of wires running through narrow tubes. When old insulation cracks, pendent lights start to fail and sparks may fly.

If one wire has brittle insulation, replace all the wires; the others are just as old.

If only one light malfunctions, turn off the switch and test its socket. Replace the socket if it is defective (page 66).

To get ready, shut off power to the circuit at the service panel. Have a helper hold the fixture, or bend the ends of a coat hanger to support it. Loosen the screws holding the canopy in place.

1 **OPEN THE FIXTURE.** Slide down the canopy. Pull out and separate the wires. Carefully remove the wire nuts. Test for the presence of power in the box (pages 44–45). If any wires are live, shut off the correct circuit. Disconnect the wires and take down the fixture. Remove the cover near the bottom of the fixture to expose the connections.

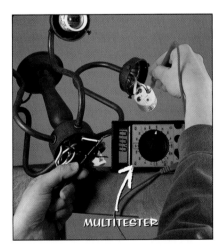

2 **TEST THE SOCKET WIRES.** If some of the sockets do not light, test each wire for continuity. To find out which cord goes where, tug on the cord at the socket end while holding the wires at the base. Unscrew the wire nuts at the base, and test both the ridged (neutral) wire and the smooth wire for continuity. If you do not get a positive reading for both wires, replace the cord. Test and repair all malfunctioning light sockets.

3 **TEST THE STEM WIRES.** If all the lights fail to come on, the stem wires probably need replacing. To make sure, twist the stem wires together at the base. Touch tester probes to both wires at the top of the fixture. If no continuity is indicated, replace the stem wires as shown on page 68.

ELECTRICAL REPAIRS

71

REPLACING PLUGS AND SWITCHES

ELECTRICAL REPAIRS

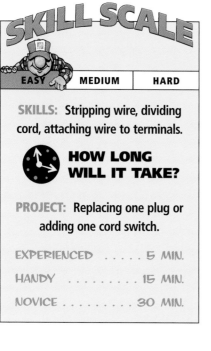

SKILL SCALE

EASY	MEDIUM	HARD

SKILLS: Stripping wire, dividing cord, attaching wire to terminals.

HOW LONG WILL IT TAKE?

PROJECT: Replacing one plug or adding one cord switch.

EXPERIENCED . . . . . 5 MIN.

HANDY . . . . 15 MIN.

NOVICE . . . . 30 MIN.

✓ STUFF YOU'LL NEED

TOOLS: Combination strippers, utility knife, screwdriver

MATERIALS: Replacement plug or cord switch

A plug with loose prongs or a cracked body is dangerous and should be replaced. If the cord and the plug are damaged, rewire the lamp or appliance with a one-piece cord and plug (page 68). If only the plug is damaged, save yourself the chore of rewiring the entire device by using one of the replacement plugs shown here.

A cord switch, which is almost as easy to install, is ideal for lamps that have hard-to-reach switches. You can add a cord switch in minutes.

INSTALLING FLAT REPLACEMENT PLUGS

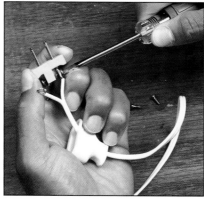

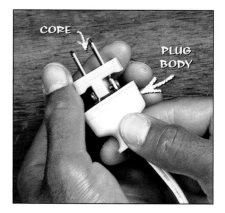

1 **JOIN THE WIRES TO A FLAT REPLACEMENT PLUG.** Cut the cord near the old plug. Slide the cord through the replacement plug body. Separate and strip the cord wires (page 69). Twist the wire strands tightly with your fingers, and wrap the strands clockwise around the core terminals. Tighten the screws.

2 **SNAP ON THE BODY.** Make sure the connections to the terminals are tight. Hold the core with one hand and push the body onto it with the other hand until the two pieces snap together.

INSTALLING A QUICK-CONNECT PLUG

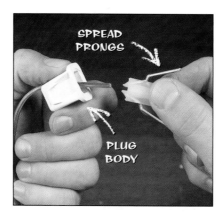

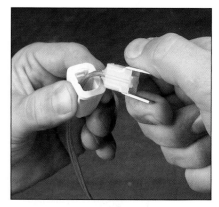

1 **INSERT THE WIRES.** With this type of plug, you do not have to divide or strip the cord. Cut off the old plug. Thread the cord through the plug body. Spread the prongs apart, and push the cord into the core. Connect the ridged (neutral) wire to the wider prong.

2 **SQUEEZE AND SLIDE TOGETHER.** Squeeze the prongs together so they bite down on the cord. While still squeezing, slip the body onto the core until it snaps into place.

INSTALLING A ROUND PLUG

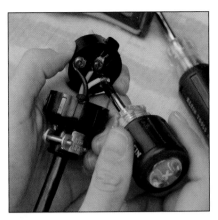

1 **STRIP THE CABLE AND INSERT THE WIRES INTO THE PLUG BODY.** If you are replacing an old plug, cut the old cord near the plug. Separate the replacement plug core and body. Use a wire stripper or a utility knife to strip 1½ inches of sheathing, being careful not to nick the insulation of the three wires inside. Strip about ½ inch of insulation from each of the wires. Thread the cord through the new plug body.

2 **MAKE THE CONNECTIONS.** Twist the wires together with your fingers so there are no loose strands. Wrap each wire clockwise around a terminal on the core: black wire to one prong, white wire to the other, and green wire to the round grounding prong. Tighten the screws, snap the body onto the core, and tighten the clamp screws to the cord to secure the plug to the cable.

SAFETY ALERT!

RATE THE WIDTH OF YOUR APPLIANCE CORDS

Most lamp cords are a standard thickness, but appliance cords vary. When you buy an appliance cord, make sure it is rated to handle the appliance amperage (page 62-63). A cord that's too thin will dangerously overheat.

To see whether a cord needs to be replaced, bend it at several points. If the insulation cracks or feels like it's about to crack, replace the cord.

ELECTRICAL REPAIRS

BUYER'S GUIDE

QUICK-INSTALL CORD SWITCHES

Choose the location of the switch carefully: You won't be able to move it after it is installed. An inexpensive **rotating switch** (below left) will not last long if the switch is used daily but is fine for occasional use. To install one, use a utility knife to cut a 1-inch-long slit to divide the cord; then snip the smooth (hot) wire, but do not strip it. Insert the wire into the switch, and screw the two halves of the switch together.

A **rocker switch** (below center) is more solidly built and will last longer. It will fit with a flat lamp cord or a round appliance cord. It takes a few minutes longer to install: Cut and strip the smooth (hot) wire, and connect the ends to terminals.

A **toe-button switch** (below right) is ideal for torchiers and plant lights that are otherwise hard to reach. It installs like on-cord rotating and rocker switches.

PUSH UNCUT WIRE INTO CHANNEL

CLIP ONE WIRE

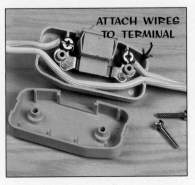

ATTACH WIRES TO TERMINAL

TESTING SWITCHES

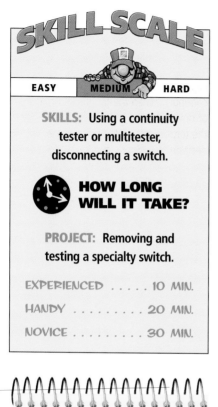

SKILL SCALE

EASY   **MEDIUM**   HARD

SKILLS: Using a continuity tester or multitester, disconnecting a switch.

HOW LONG WILL IT TAKE?

PROJECT: Removing and testing a specialty switch.

EXPERIENCED . . . . . 10 MIN.

HANDY . . . . 20 MIN.

NOVICE . . . . 30 MIN.

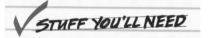

✓ STUFF YOU'LL NEED

TOOLS: Continuity tester or multitester, screwdriver

MATERIALS: None needed

A switch should show continuity when turned on and no continuity when turned off. With some specialty switches, however, it might be hard to know when the switch is on and when it is off. Some of the more common specialty switches are expensive enough to warrant testing before you replace them. (You may want to replace single-pole switches without testing—they're cheap to replace.) Do not test a switch while it is wired. Shut off power to the circuit, and remove the switch.

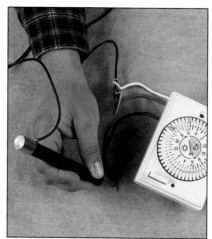

TESTING A TIMER SWITCH. Turn the dial until the red ON tab just passes the indicator arrow and clicks on. Touch one probe to the red lead and the other probe to the black lead. If there is no continuity, the switch is defective; replace it. Also turn the dial until the OFF tab passes the arrow and clicks off. Touch the red and black leads again. If the tester shows continuity, the switch is defective; replace it.

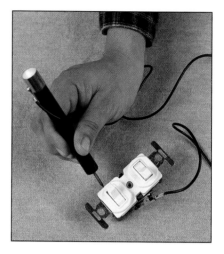

TESTING A DOUBLE SWITCH. Test each of the switches in the same way: Touch the probes to the terminals on each side of the switch. If the tester indicates continuity with the switch ON and no continuity with the switch OFF, then the switch is working. If you get any other result from either switch, replace the device.

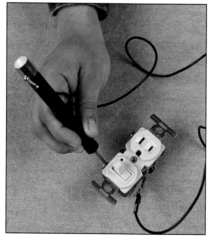

TESTING A SWITCH/RECEPTACLE. Begin by testing the switch. With probes touching the terminal on each side, the tester should show continuity with the switch ON and no continuity with the switch OFF. If you get different results, replace the device.

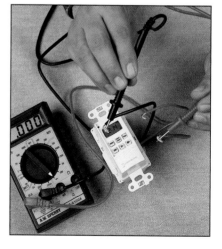

TESTING A PROGRAMMABLE SWITCH. Turn the manual override switch to ON, and touch the probes to both leads. Use a digital multitester as shown or a continuity tester (pages 44–45) to test for continuity. Then test with the switch turned OFF. The tester should show no continuity. If your results differ, replace the switch.

TESTING A THREE-WAY SWITCH

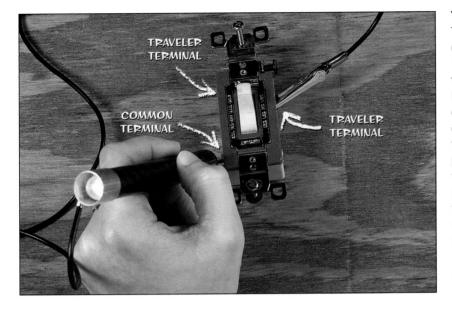

TESTING A THREE-WAY SWITCH.
Touch one probe to the common terminal (it is a different color and may have "common" printed next to it) and one to either of the other "traveler" terminals. Flip the switch. The tester should show continuity when the toggle is either up or down, but not in both positions. Keep the toggle in the ON position (the position that shows continuity) for the first traveler terminal, and move one probe from the first traveler terminal to the second. The tester should show no continuity. Flip the switch, and the tester should show continuity. If any of the test results differ, replace the switch. (See pages 148–150 for more about three-way switches.)

UNDERSTANDING AND TESTING A FOUR-WAY SWITCH

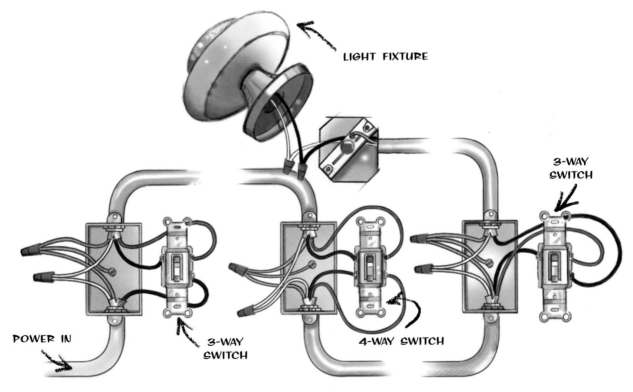

LIGHT FIXTURE

3-WAY SWITCH

POWER IN

3-WAY SWITCH

4-WAY SWITCH

A FOUR-WAY SYSTEM CONTROLS A SINGLE LIGHT WITH THREE OR MORE SWITCHES. The first and last switches are three-ways, and the switch or switches between them are four-ways. Carefully tag all the wires before removing any of the switches. This schematic will help if you get confused. Test the three-way switches as described above, but you'll have to take the four-way to an electrical supply store for testing. What's so complicated about a four-way? The switch should have four paths for continuity. The direction of the paths depends on the manufacturer that made the switch. The paths of continuity may run crosswise or diagonally from any of the four terminals to any of the others.

75

CHECKING 240-VOLT RECEPTACLES

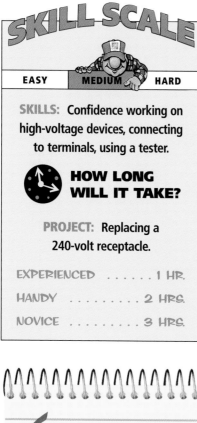

✓ STUFF YOU'LL NEED

TOOLS: Screwdriver, continuity tester or multitester, longnose pliers

MATERIALS: None needed

SAFETY ALERT!

TAKE EXTRA PRECAUTIONS
240 volts can do a lot more damage than 120. When doing a live test, wear rubber-soled shoes and, if kneeling, use a rubber pad. Do not touch the metal parts of the tester probes. (See pages 5–6 for other important safety tips.)

Stationary 240-volt appliances, such as electric water heaters, central air-conditioning units, and electric furnaces, are "hardwired." Instead of having cords with plugs, a cable runs directly from the appliance to a junction box.

Movable 240-volt appliances, such as window air conditioners and electric ranges, are plugged into 240-volt receptacles.

Some receptacles deliver both 240-volt and 120-volt power (page 145). These are used for electric ranges and clothes dryers that need heavy voltage for heating elements and standard voltage for motors and clocks.

Specific types of receptacles are available, each with a different hole configuration so only one type of plug can fit. Ranges usually use 120/240-volt, 50-amp receptacles; dryers plug into 120/240-volt, 30-amp receptacles; air conditioners use 240-volt, 30-amp receptacles.

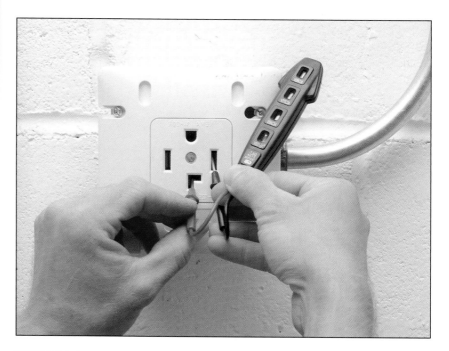

TESTING A 240-VOLT RECEPTACLE. If an appliance plugged into a 240-volt receptacle gets no power or only partial power, first check the service panel to make sure that the breaker hasn't tripped or the fuse hasn't blown. To make a live test, turn on the circuit, and carefully insert the probes of a four-level voltage tester or a multitester (pages 44–45) into the two vertical slots. The meter should register around 240 volts. With one probe in a vertical slot and one in a neutral or ground slot, you should get a reading of 120 volts. If your readings differ, shut off power to the circuit and remove the receptacle. Make sure the wiring connections are tight. If they are not, tighten and retest. Otherwise, replace the receptacle with a duplicate (page 145). If the receptacle is working correctly, but the appliance is not, you may need to replace the appliance cord.

SWITCHING FROM THREE-WIRE TO FOUR-WIRE

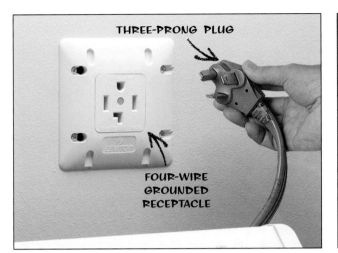

THREE-PRONG PLUG

FOUR-WIRE
GROUNDED
RECEPTACLE

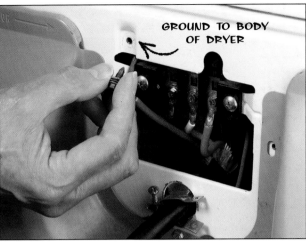

GROUND TO BODY
OF DRYER

1 **SWITCHING FROM A THREE-WIRE TO A FOUR-WIRE DRYER RECEPTACLE AND CORD.** In many areas, codes require that electric dryer receptacles have four holes connected to four wires—two hot, one neutral, and one ground. But you may have a four-hole receptacle and an old dryer that has a three-prong plug. If so, replace the dryer cord and plug. Buy the correct type of cord with plug at a home center or electrical supply store. Unplug the dryer. Open the access panel on the back of the dryer.

2 **ATTACH THE WIRES.** Kneel on a foam pad while you work. Note how the three-wire cord is attached, and wire the new cord the same way. A ground for the dryer motor often attaches to the dryer body near where the plug wires attach. Attach the cord's ground wire to it.

MOVE UP TO A GROUNDED RECEPTACLE

If your system is grounded (look for a copper wire fastened to the box and check for grounding at your service panel), give yourself a safety edge by installing a four-wire receptacle. Attach black to black, white to white, and red to red. With #10 wire, pigtail to the ground. If you need to run new cable for a new receptacle, see pages 121-132.

STRAIN-RELIEF
BRACKET

3 **TIGHTEN THE STRAIN-RELIEF BRACKET.** Don't neglect this important piece of hardware; if the dryer cord gets yanked, the bracket will protect the connections and help avoid a possible short. Fit it into the cord access hole and evenly tighten both screws firmly onto the cord.

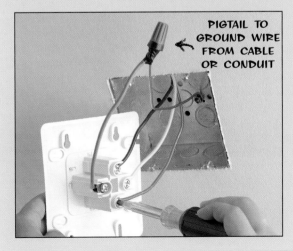

PIGTAIL TO
GROUND WIRE
FROM CABLE
OR CONDUIT

REPAIRING FLUORESCENTS

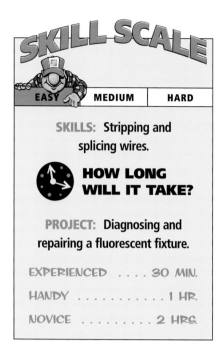

SKILL SCALE

EASY	MEDIUM	HARD

SKILLS: Stripping and splicing wires.

HOW LONG WILL IT TAKE?

PROJECT: Diagnosing and repairing a fluorescent fixture.

EXPERIENCED . . . . 30 MIN.

HANDY . . . . 1 HR.

NOVICE . . . . 2 HRS.

STUFF YOU'LL NEED

TOOLS: Combination strippers, longnose pliers, screwdriver

MATERIALS: Replacement parts, electrician's tape, wire nuts

Fluorescent lights use less energy and last longer than incandescent lights, but they can be finicky to repair. The greatest challenge can be finding the right replacement parts. Starters will need replacing on older fixtures (newer fixtures don't need them). Consider replacing starters when you replace tubes. Sockets can loosen or crack; ballasts are particularly troublesome and expensive to replace. Save yourself time and trouble by taking down the fixture to repair it on a bench.

TROUBLESHOOTING

A **flickering** or **partially lighted** tube is the most common problem. Take these steps to troubleshoot:
- Rotate it for a better connection.
- Replace the starter.
- Replace the ballast or the fixture.

If a tube has **very dark spots** at either end:
- Replace it, even if it works. It may cause the ballast to wear out.

If the tube **does not light** at all:
- Rotate the tube to get a better connection.
- Check the ballast for a temperature rating. Some fixtures will not start in cold or hot temperatures.

- Make sure the circuit is getting power. Test, then replace the wall switch (pages 22–23) if necessary.
- Replace the tube, especially if it's dark at the ends or if a pin is bent.
- Replace the socket if it is cracked or if the tube does not seat tightly.
- Replace the ballast or the fixture.

If the **ballast hums:**
- Try turning off a nearby radio or heavy-use electrical appliance.
- Tighten the ballast-mounting screws.
- Replace the ballast.

If the ballast is **oozing a thick black substance:**
- Replace the fixture or, wearing protective gloves, replace the ballast.

CLOSER LOOK

KNOW YOUR FIXTURE

It takes an initial burst of voltage to light a fluorescent tube. Once it's lighted, current is cut back because the tube can "coast" on very little voltage. The ballast, a transformer, steps down the voltage. In older models, the ballast is a bulky and heavy rectangular object. Newer models have electronic ballasts. In a rapid-start fixture (below left), the ballast performs this two-level delivery of power. In a starter-type fixture (below right), a small cylindrical starter acts as a switch, sending a greater amount of current to the tube until it lights.

BALLAST

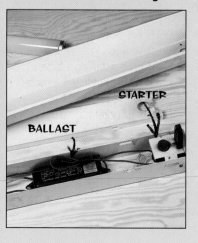

STARTER

BALLAST

IDENTIFYING SIGNS OF TUBE WEAR. If a tube suddenly stops lighting and is not blackened at the ends, gently rotate it while the fixture is turned on and see whether that brings it back to life. Gray spots near the ends of a tube (top) are signs of normal aging. If the ends are black or dark gray (bottom), you should replace the tube. If a fixture has two tubes, always replace both at the same time.

REPLACING A TUBE. To remove a tube, hold it at each end and twist carefully until you feel it loosen. Remove it, being careful not to damage the tube pins or the sockets. Replace it with a tube of the same size and wattage. With dual lamps, replace both bulbs at the same time.

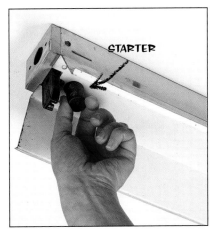

REPLACING A STARTER. If it takes more than a few seconds for a starter-type fixture to light up, remove the tube and twist the starter to see whether you can seat the starter more firmly. If the ends of a tube light up but the center doesn't, replace the starter. Press in the starter and twist counterclockwise to remove it. Buy a starter with the same part number as the old one. Push in and twist clockwise to install it.

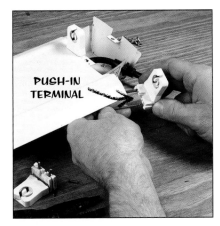

REPLACING A SOCKET. These crack easily, especially if you are not careful when removing or installing a tube. Unscrew the bracket holding the socket in place, or slide the socket out of the groove. If the socket has push-in terminals, poke the slot to release the wire. If the socket has attached wires, cut the wires and strip off about ½ inch of insulation. Install a new socket with push-in terminals or screw terminals.

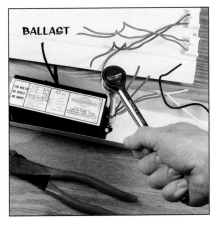

REPLACING A BALLAST. Shut off power to the circuit supplying the light, and check for the presence of power. Disconnect the wires if possible. If it's not possible, cut them close to the ballast. Either way, tag the remaining wires so you'll remember which wire goes where. Unscrew the ballast, and take it to a home center or electrical supply store for a replacement. Install the new ballast in just the same way as the old one was installed. You may prefer to replace the fixture entirely.

Homer's Hindsight

EASY DOES IT
While installing a replacement socket for my fluorescent fixture, I poked the wires into the push-in terminals as hard as I could—the harder the better the connection, right? Wrong. I pushed the wires in so far that the insulated part (rather than the stripped part) of the wire connected to the terminal. No connection, no light! Remember: Just push until you feel the clip inside the socket grab the stripped wire.

RESETTING BREAKERS

ELECTRICAL REPAIRS

The fuses or circuit breakers in the service panel form the first line of defense for your home, protecting you and your family from fire and shock. If a house is wired correctly, with no circuits overloaded, you may never have to open your service panel except to shut off power while working on an electrical project.

If a circuit in your home frequently blows a fuse or trips a breaker, check pages 62–63 for tips on how to eliminate circuit overloads.

Learn how to shut off and restore power from the service panel. Map your circuits and tape an index in your service panel (page 61). Always leave a clear pathway to the service panel.

If a circuit breaker trips often, even though you don't seem to be running too many appliances or lights, the problem may be the wiring or the circuit breaker. It's easy to test a breaker (see below).

CLOSER LOOK

HOW BREAKERS TRIP
Service panels and breakers are made by a number of manufacturers, so there are various ways to reset breakers. Here are some common types of breakers.

This type flips halfway toward OFF when it trips. To reset it, turn it off, then on.

This breaker flips off all the way. Just flip it back on to reset it. On some, a red button displays or pops out, showing that the breaker has tripped.

This breaker model has a button that pops out when it trips. Push the button in to reset.

TESTING A BREAKER

1 **TEST THE CIRCUIT BREAKER.** If you suspect that a faulty breaker is tripping for no apparent reason, touch the prongs of a voltage tester to the breaker's terminal screw and a ground. If there is no power, the breaker is faulty. Or try this test: Shut off the main breaker. Loosen the setscrews on the suspected breaker and a nearby breaker of the same amperage. Switch the wires, tighten the setscrews, and flip the main breaker back on. If the original breaker trips unreasonably while connected to a different circuit, replace the breaker.

2 **REPLACE THE BREAKER.** Shut off the main breaker to be safe. Loosen the setscrew on the damaged breaker, then pull out the wire. Pull out the breaker by hand. Make sure you touch only plastic, never anything metal. Pull out one end of the breaker to loosen it, and then pull out the whole breaker. Buy a new breaker of the same amperage and size, made by the same manufacturer. Slip the wire into the new breaker and tighten the setscrew. Push the breaker in until it snaps in place like the ones around it. Restore power.

CHANGING FUSES

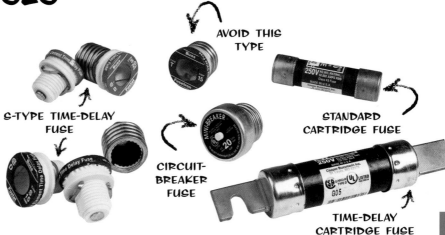

S-TYPE TIME-DELAY FUSE

AVOID THIS TYPE

CIRCUIT-BREAKER FUSE

STANDARD CARTRIDGE FUSE

TIME-DELAY CARTRIDGE FUSE

Always replace a blown fuse with a fuse of amperage appropriate for the circuit. A living area usually requires a 15-amp fuse; an appliance area needs a 20-amp fuse. A 30-amp fuse is used only for a range or dryer circuit, or for a line to a subpanel. Installing a fuse of higher amperage may get the circuit going again, but it puts your house at risk because the fuse won't blow when wires get dangerously hot.

TYPES OF FUSES

A **time-delay fuse** holds itself together for a second or so during a momentary surge of power—for example, when a refrigerator motor turns on. The fuse will blow if the circuit remains overloaded.

An **S-type** fuse has a socket adapter that screws into the fuse box socket, where it becomes permanently lodged. Once screwed in, it is impossible to install a fuse of a different amperage.

A **circuit-breaker fuse** has a push button that pops out when the circuit overloads. Instead of replacing the fuse, you push the button back in to restore power. Many electricians don't think they're reliable; others consider them safe.

WHY A FUSE BLEW

If the metal strip inside the fuse is broken completely, the circuit overloaded: Too many appliances and lights were running at the same time. If the fuse window is

blackened, the cause is a short circuit—meaning that somewhere wires are touching each other or a wire is making contact with metal. Inspect switches, receptacles, and fixtures—and fix the problem right away.

SHORT   **OVERLOAD**

WORKING WITH CARTRIDGE FUSES

1 **REMOVE THE FUSE FROM THE BLOCK.** If a 240-volt circuit in a fuse box blows, the fuses are probably located inside a fuse block. Turn off the power. Grab the wire handle and pull out the block. Use a fuse puller to remove each cartridge fuse.

2 **TEST THE CARTRIDGE FUSE.** (Be careful. If you have just removed the fuse, its metal parts may be hot.) To see whether a cartridge fuse has failed, touch both ends with the probes of a continuity tester or multitester (pages

44–45). If the fuse tests positive for continuity, it is good. If not, it has blown. Take it to a home center or hardware store and buy an exact replacement.

REPAIRING WIRES IN BOXES

SKILL SCALE

EASY	MEDIUM	HARD

SKILLS: Wrapping tape around wires in tight spots.

HOW LONG WILL IT TAKE?

PROJECT: Taking measures to safeguard several wires in a box.

EXPERIENCED . . . . . 10 MIN.

HANDY . . . . 30 MIN.

NOVICE . . . . 45 MIN.

✔ STUFF YOU'LL NEED

TOOLS: Screwdriver

MATERIALS: Electrician's tape, BX bushings

ELECTRICAL REPAIRS

You open a box in your older home and find old wiring with insulation that is cracked and frayed. Very likely, all the hidden wires in the house are in equally bad shape. What can you do?

Rewiring is the safest solution. It is not too big a job if all the wires run through conduit or Greenfield flexible conduit (pages 126–127), but many homes are wired with cable. Replacing cable means making holes in walls, followed by time-consuming, expensive patching and redecorating.

Wires wrapped tightly in cable are likely to be in better condition than wires exposed to air. Insert a plastic bushing and tape the wires to protect the circuit until you rewire. Better yet, protect the wire with a hot-shrink sleeve (above right).

While the box is open, take the following precautions as well. A box recessed behind the wall surface poses a fire danger and is out of code. Slip in an extender ring (below right), and add the cover plate. Debris that collects in an electrical box, especially sawdust, poses a fire hazard; vacuum it out immediately.

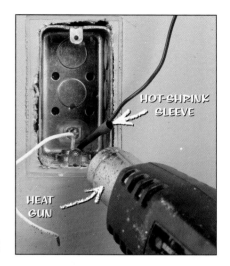

PROTECT A WIRE WITH A HEAT-SHRINK SLEEVE. If a wire has cracked, brittle, or otherwise damaged insulation, buy a small bag of plastic sleeves made to protect wires. Shut off power to the circuit. Disconnect the damaged wire, and slip a sleeve down over it. Point a hair dryer or heat gun at the sleeve until it shrinks, forming a long-lasting protective coating.

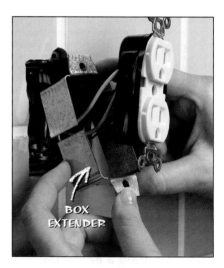

EXTEND A RECESSED BOX. If a box is recessed from a wall surface, there is a danger of fire—especially if the wall surface is wood paneling. However, even tile edges or drywall should be covered. Purchase a box extender sized to fit your box, and slip it on.

CLOSER LOOK

LEAKY WIRES

Wiring with cracked insulation can leak small amounts of electricity. Known as a high-resistance short circuit, this power loss won't blow a fuse or trip a breaker, but it can overheat wires. To test for this problem, completely shut down the house. Remove all lightbulbs, unplug all lamps and appliances, and disconnect hardwired appliances such as electric water heaters and whole-house fans. Turn on all the switches. Then watch your electric meter. If it shows power usage, then you have a high-resistance short.

Test circuits one by one to narrow down the source of the leak. It may be a bad connection or damaged wire insulation in an electrical box. If you can't find the source, call in a professional.

6 PLANNING LIGHTING

Before you choose light fixtures, draw up a lighting plan. Keep in mind that lighting does more than just illuminate. Lighting also:

- **Enhances activities.** Reading, food preparation—even cleaning—is easier and more pleasurable when the lighting is ample but not glaring.

- **Highlights decorative features.** By changing or redirecting lights, you can emphasize artwork, favorite pieces of furniture, or other decorative features. You can even position lights to make a room seem larger. Outdoor lighting can dramatize built-ins and plantings.

- **Sets a mood.** By building versatility into a room with a variety of fixtures and dimmer switches, you can easily adjust the lighting to suit the occasion.

- **Provides safety.** Well-placed lights help make stairs and hallways safer; outdoors, lighting can even discourage intruders.

CHAPTER SIX PROJECTS

GOOD IDEA

ENERGY EFFICIENT LIGHTING

Consider replacing regular incandescent bulbs with energy-efficient fluorescents. They last longer and produce the same quality of light.

CHOOSING CEILING FIXTURES

The broad range of overhead fixtures can be roughly divided into ones with eye-catching decorative features (like the ones shown on this page) and ones that are hardly noticeable but provide general illumination (like the flush ceiling fixtures shown opposite below). Track lights (opposite) fall in between. All come in a wide variety of styles. Here are the basic types and features to choose from.

PENDANT LIGHTS

Lights that hang down from the ceiling are called pendants. Use them for general lighting, to illuminate a dining room table, or to light up a work surface.

A chandelier or other type of pendant usually can't illuminate a large room on its own. That's because a chandelier often hangs at eye level and would produce an unpleasant glare if it were bright enough to light an entire room.

■ **Pendant shades.** Use a pendant shade to focus light on a specific space, such as a small table, a countertop, or a narrow work area. A pendant light with a glass shade will provide general lighting as well as directed light. A metal shade focuses light more directly. Older styles of pendant lights hang by decorative brass chains, with neutral-colored lamp cord running through the chain. Newer fixtures use a plain chrome-colored wire for support, with the cord running alongside.

■ **Pendant lanterns.** These lights resemble the old glass lanterns that protected candles from wind. Use them in narrow areas like foyers and stairways. Hang these at least 6½ feet from the ground so that people can walk under them. Center a pendant lantern width-wise in a narrow room. If it is near a large window, place it so it will look centered from the outside.

■ **Chandeliers.** Originally designed as candleholders, chandeliers usually have five or more light bulbs. Look for a model that is easy to clean; complex designs can be difficult to dust. Keep it in scale—a chandelier that is too small will appear to be dwarfed by the room. When choosing a unit to hang over a dining room table, select one that is about 12 inches narrower than the table. If it is any wider, people may bump their heads on it when they stand up from the table.

PENDANT
LANTERN

PENDANT
SHADE

CHANDELIER

In an entryway, maintain proportion by installing a chandelier that is 2 inches wide for every foot of room width—for example, use a 20-inch-wide light in a 10-foot-wide room.

Get the height right. A common mistake is to hang a chandelier too low. A chandelier should hang about 30 inches above a table top. The length of the chain will depend on your ceiling height.

TRACK CEILING FIXTURES

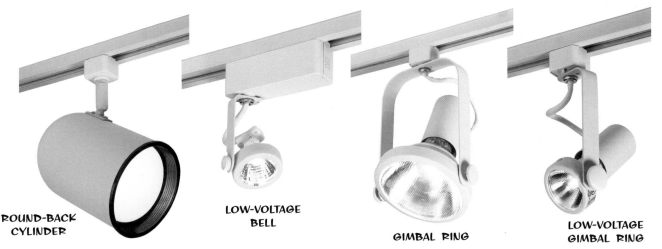

ROUND-BACK CYLINDER

LOW-VOLTAGE BELL

GIMBAL RING

LOW-VOLTAGE GIMBAL RING

CHOOSING TRACK LIGHTS. A single track lighting system can combine general lighting and accent lighting. When choosing a lamp, make sure it can handle the light bulb of your choice and that it will fit onto your track. Incandescent lamps such as a **round-back cylinder** or a **gimbal ring** produce a broad, intense beam. Low-voltage halogen track lights such as a **low-voltage bell** or **low-voltage gimbal ring** produce a more intense, less broad area of light. They have their own transformer, so they can attach to a standard-voltage track. (However, these low-voltage lights require a special dimmer switch; a standard dimmer will damage the lamps.) A track that partially encircles a room at a distance of 6 feet or so from the walls will disperse light more effectively than a single track running through the middle of the room.

FLUSH CEILING FIXTURES

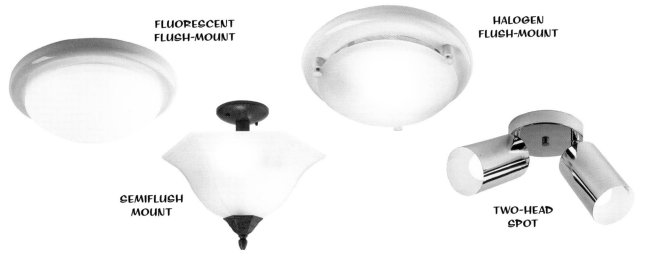

FLUORESCENT FLUSH-MOUNT

HALOGEN FLUSH-MOUNT

SEMIFLUSH MOUNT

TWO-HEAD SPOT

CHOOSING FLUSH FIXTURES. A **single flush fixture** in the middle of the ceiling is the most common way to light a room. These fixtures usually produce enough light to adequately illuminate a 12×12-foot room with an 8-foot ceiling or a 16×16-foot room with a 10-foot ceiling (the higher the fixture, the broader the spread of its light). They hug the ceiling, consistently distributing light. Newer **fluorescent ceiling fixtures** with electronic ballasts look like incandescents, save energy, and have tubes that rarely burn out. A **semiflush** fixture hangs down a foot or so from the ceiling. It diffuses light through the globe as well as upward like a cove light, evenly illuminating a room. **Halogens** offer more intense light. **Two- or three-head spotlights** provide some of track lighting's versatility. Point the lights horizontally for general lighting, or angle them downward to highlight certain areas of the room.

SELECTING BULBS AND TUBES

The color of a light bulb or a light fixture globe or shade significantly affects the mood of a room. Lighting that is slightly red or yellow is considered "warm," while blue-tinged light is "cool." Incandescent bulbs produce warm light; many fluorescents are cool—if not downright cold.

Choose the color of your home's lighting according to the color of your furnishings. If you have pure white walls or cabinetry, warm lighting will make them beige. Cool light directed at brownish natural wood may give it a green tinge.

Fortunately, whether you have a fluorescent or an incandescent fixture, you can switch from cool to warm light, or vice versa, by changing the bulbs or tubes.

BUYER'S GUIDE

NEW LIGHT ON FLUORESCENTS

Fluorescent lighting is economical but often harsh and cold. For a slightly higher cost, you can buy tubes that deliver light similar to that of an afternoon sun. The lower the Kelvin temperature of a tube, the warmer its light will be. A tube marked "3000K," for example, delivers warm color, while a "5000K" tube will make a room feel cool. "Full-spectrum" or "wide-spectrum" tubes have low Kelvin ratings.

SUNSHINE OR DAYLIGHT

AQUARIUM/PLANT

WARMER THAN COOL BUT COOLER THAN AQUARIUM

COOL

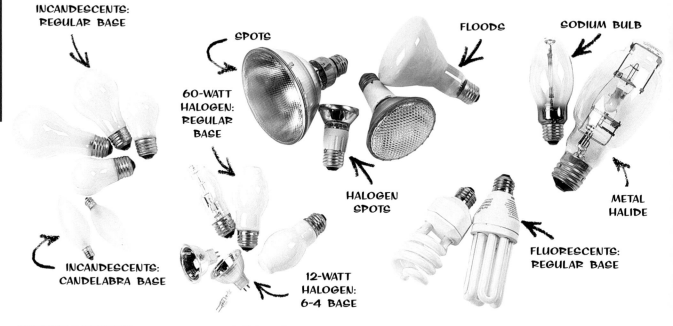

INCANDESCENTS: REGULAR BASE

SPOTS

60-WATT HALOGEN: REGULAR BASE

HALOGEN SPOTS

FLOODS

SODIUM BULB

METAL HALIDE

FLUORESCENTS: REGULAR BASE

INCANDESCENTS: CANDELABRA BASE

12-WATT HALOGEN: 6-4 BASE

LIGHTBULB OPTIONS.

■ **Incandescent bulbs** are the most common but have comparatively short lives and are not very energy efficient.

■ **Low-voltage halogen bulbs** last longer than incandescents and use far less energy, but they burn hot. Halogens come in so many styles, so make sure the bulb base fits in your fixture.

■ **Reflector bulbs** direct either a wide or narrow beam of light, depending on the bulb. A "spot" bulb projects a flashlight-like beam. A "flood" bulb illuminates a wider area. The second number on the stamped label indicates the degree of the beam spread.

■ **Fluorescent tubes** that screw into incandescent sockets save money in the long run. Choose from among several shapes and degrees of warmth.

■ **HID (High-Intensity Discharge) lamps** such as sodium, metal halide, and mercury vapor produce very bright, economical light outdoors.

PLANNING KITCHEN LIGHTING

The right lighting plan can make your kitchen more cheerful and inviting, increase the safety of food preparation, and highlight cabinetry and other design features. As you plan, remember that surfaces like ceramic tiles and semigloss paint reflect light. This can be a beneficial, but in the wrong place they can bounce bright light into your eyes.

■ **Ambient lighting** produces a daylight effect. Flush ceiling fixtures or track lights spread light more evenly than recessed can lights or pendants. Cove lighting creates ambient light originating from several directions. Windows and skylights are great sources of light during daylight hours, but they need help in the evening or in gloomy weather. A dimmer switch or two on your ambient lighting will make it easier to strike the right balance.

■ **Task lights** under kitchen cabinets or in other strategic areas illuminate common kitchen tasks like food preparation and dish washing. A range hood with a light eases stove-top cooking as it vents odors.

■ **In-between lights** illuminate kitchen work spaces while providing generous amounts of ambient light. These lights include recessed can lights over a sink, pendant fixtures above an eating area, and track lights in a semicircle near cabinetry.

AMBIENT LIGHTING

TASK LIGHTS

IN-BETWEEN LIGHTS

TRACK LIGHTING

PENDANT TASK LIGHTING

FLUORESCENT LIGHT OVER WORK AREA

ROPE LIGHTING

LIGHTING YOUR EATING AND PREP AREAS.
Ensure that ambient lighting is positioned so that it amply illuminates work areas. To supplement ambient light, install fluorescent or under-cabinet halogen lights over work surfaces. If there are no cabinets above, use track lighting, sconces, recessed can lights in the ceiling, or halogen trapeze lights. Pendant lights work well for task lighting but are not practical above a sink because they hang down too low.

Strings of rope lights placed along the kickplate add an accent and highlight your flooring.

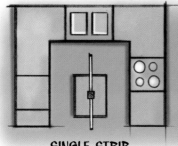

SINGLE STRIP

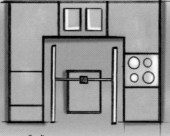

"H" CONFIGURATION

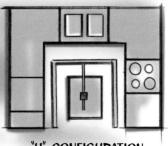

"U" CONFIGURATION

SHAPING UP WITH TRACK LIGHTING.
Many kitchens feature a single strip of track lighting running through the center of the ceiling. This kind of light provides adequate illumination but can sometimes bounce off wall cabinets and produce an uncomfortable glare—especially if the cabinets are shiny or light in color. The lights can cast a shadow over a person preparing food at the countertop, contributing to poor visibility.

Instead of installing a single strip of track lighting along the ceiling, wrap the tracks around the room in an "H" or a "U" pattern. Install the tracks about 3 feet out from the wall and 2 feet out from the wall cabinets. The lamps will then shine down over the shoulders of people working at counters, or toward the center of the room—providing both task lighting and ambient light.

LIGHTING COUNTERTOPS. Place fluorescent or halogen under-cabinet lights so they will illuminate the countertop but not shine in a person's eyes. If the light fixtures are chunky, consider installing a 2-inch strip of wood along the underside of the cabinet to shield the glare.

USING COVE LIGHTING. This is an easy and inexpensive way to add an elegant lighting touch to a kitchen. Fluorescent fixtures placed on the top of a wall cabinet wash the wall and ceiling in a glow that disperses even light throughout the kitchen.

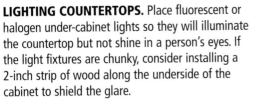

LIGHTING FOR HEALTHY PLANTS

SUPPLEMENTING NATURAL LIGHT. Nothing brightens a kitchen window like potted flowers or herbs. Most plants need at least 4 hours of direct sunlight a day. Unless your window provides this kind of exposure, you may need to supplement your room's natural light with artificial rays. Install incandescent or fluorescent bulbs or tubes that are labeled full- or wide-spectrum. As supplements to filtered window light, these lights only need to be on for several hours a day. Keep them a foot or more away from your plants to avoid drying out the leaves.

If your room provides little natural sunlight or none at all, you will have to install supplemental lighting at close range. To grow healthy plants, train these lights on your plants and leave them on for all or most of the day.

LIGHTING A BATHROOM

An average-size bathroom needs a ceiling fan/light in the center of the main room, a moisture-proof ceiling light over the shower/bath, and lights over the sink.

■ **Ambient lighting** is typically provided by an overhead light combined with a vent fan. Make sure the fan's blower is powerful enough to adequately vent your bathroom (pages 156–158). For a little more money, you can also purchase a low-watt night-light or a fan-forced heater unit. Some people prefer a heat lamp near the tub or shower for additional heat while drying off after bathing.

■ **Bathroom mirror lighting** deserves careful thought. A horizontal strip of decorative light bulbs above the mirror provides lots of light but may shine in your eyes. A fluorescent fixture with a lens provides more even light but may lack warmth. Sconce lights placed on either side of the mirror are the best source for lighting your face for shaving or applying makeup. When planning circuits, don't forget to install a Ground-Fault Circuit Interrupter (GFCI) receptacle near the sink.

■ **Shower lighting supplements** what little light comes through the shower curtain or glass door. Consider installing a recessed canister light with a watertight lens placed directly above the shower.

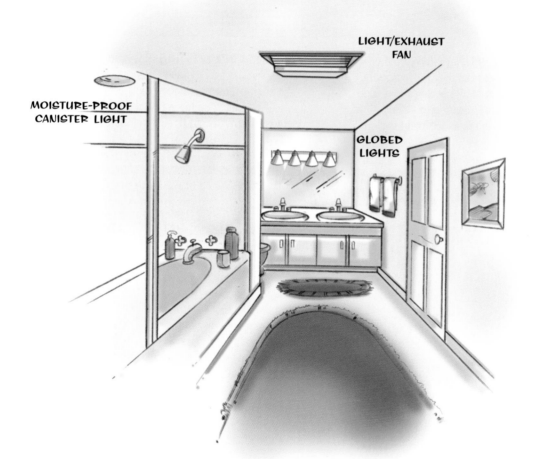

LIGHTING UP YOUR BATHROOM. The darker the color of your bathroom walls and fixtures, the more light you need. Natural light from a window may be sufficient for daytime use. But for nighttime and early-morning use, the shower, in particular, might need one or two moisture-proof canister lights. (Codes limit them to 60 watts each if the shower is enclosed.) Above the sink, install moisture-resistant globed lights that won't shine in your eyes. Overhead, install a single fixture that efficiently and stylishly combines a light, exhaust fan, and perhaps a night-light and a heater.

LIGHTING LIVING AREAS

living rooms, dining rooms, great rooms, and large bedrooms all benefit from both ambient and task lighting. Rather than installing a single lighting component, think in terms of the total effect of the room. "Layering" several types of lights makes a room more comforting and inviting. The goal is flexibility, so you can set a variety of moods by brightening or dimming the entire room or part of the room.

- **Highlight a piece of art or cabinetry** or accentuate wall texture with lights to give the room warmth and interest.
- **Put at least one of the components on a dimmer switch,** and install several lights that are optional, but not necessary. Don't be afraid to install too many lights; you don't have to have all of them on at the same time.
- **Install an in-between light** such as a dining-area chandelier to brighten the dinner table and provide some ambient light.

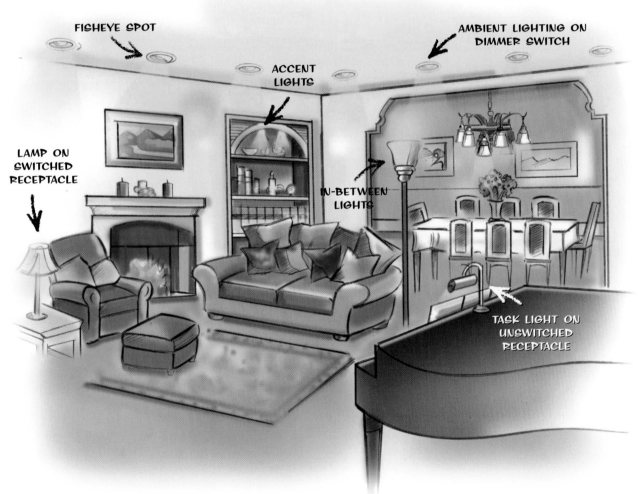

FISHEYE SPOT

ACCENT LIGHTS

AMBIENT LIGHTING ON DIMMER SWITCH

LAMP ON SWITCHED RECEPTACLE

IN-BETWEEN LIGHTS

TASK LIGHT ON UNSWITCHED RECEPTACLE

SHOWING OFF A GREAT ROOM WITH GREAT LIGHTING. The lighting plan for this large family room includes a grid of recessed canister lights for general lighting and a centrally located chandelier over the dining table. A recessed light with fisheye trim spotlights a wall painting. The table lamp and floor lamp are controlled by wall switches. Accent lights brighten shelves. The task light on the piano is on an unswitched receptacle.

LIGHTING UP CABINETS. Achieve a stunning effect with lights placed inside glass-doored cabinets. If the shelves are also glass, the light will glimmer as it filters down. Use small fluorescent fixtures—which stay cool—and control them with a special fluorescent dimmer switch.

CHOOSING A BEDSIDE LIGHT. For bedtime reading, a swivel light that mounts on a wall has great advantages over a table lamp. You can point the light directly at your book, and it won't take up space on the end table.

INSTALLING SCONCE LIGHTING. Lights that mount on the wall can make a room feel larger and a hallway wider. Use sconces for accents rather than for ambient lighting. Place low-wattage bulbs in them, unless you want to highlight the wall above.

LIGHTING ROOM BY ROOM

Provide very bright light in areas used for work and study. In rooms designed for entertaining, less light is called for.

- **Eating/dining areas.** A bright pendant light is appropriate for a breakfast nook or other informal eating area, but a table used for fine dining should have indirect, subtle light. Point recessed or track lights away from the table. Position a chandelier so it does not shine in diners' eyes.
- **Hallways and stairways.** These areas require enough light so people won't trip. You may need lights at the bottom and top of a stairway. A 75-watt ceiling fixture every 12 feet is sufficient for a hallway. Increase the wattage if elderly people live in your home. Wall sconces work well in these areas.
- **Study.** A single reading lamp can create an uncomfortable glare on book pages. Provide one or two additional sources of light, such as ample overhead lighting or a second lamp.
- **Work rooms/hobby areas.** Start with overhead lighting that distributes light evenly throughout the room. Then add nonglare lights above work surfaces and flexible lamps for specific tasks.

PLANNING FOR CAN LIGHTING

an lights vary in intensity and angle. The higher your ceiling, the more floor space a light will illuminate. In general, recessed cans should be 6 feet from each other. Of course, most rooms are not sized to accommodate this, so you'll have to adjust your calculations. In the example below, most of the lights are 5 feet apart. Make a similar plan for your own installation, experimenting with several configurations. Take your plans with you to your home center for advice.

Once you start installing can lights, you'll find that many have to be moved several inches from their ideal locations in order to avoid hitting joists (ceiling framing). Fortunately, this will not make a big difference in the overall effect.

SPECIAL TECHNIQUES

In addition to providing general lighting, recessed can lights can enhance decorating strategies with:

- **Wall washing.** To light up a large wall area, install cans with wall-wash trims that are 24 to 30 inches apart, and the same distance from the wall.
- **Accent lighting.** Spotlight a painting, fireplace mantle, or other feature with a can that has an eyeball trim. Place it 18 to 24 inches from the wall, centered on the object.
- **Grazing.** To dramatize an unusual vertical surface, such as a fireplace or a textured wall, place cans 6 to 12 inches from the wall and 12 to 18 inches apart. Wire them with a separate dimmer switch.

CONCENTRIC CIRCLES OF LIGHT. On graph paper, make a scale drawing of your room and place dots in a fairly consistent pattern. To get a general idea of the distribution of light, use a compass to draw circles that are scaled to about 5 feet in radius (10 feet in diameter). In this example, the center of the room will get more light than the perimeter—which is usually desirable. Generally, figure that a 65-watt floodlight in a room with an 8-foot ceiling will light up a circle that is 8 feet in diameter; if the ceiling is 10 feet high, it will illuminate a 10-foot circle.

PLANNING SECURITY LIGHTING

(page 99)

Outdoor lighting may be your home's most important security feature. It may even deter intruders more effectively than additional door locks or an alarm system.

- **Keep areas brightly lit** so there are no dark pathways. Ideally, two or more lights should be pointed at a potential intruder who approaches your home.
- **Install two light fixtures at each door of entry**—or at least one fixture equipped with two bulbs in case one bulb burns out. Control these lights with motion sensors or timers (page 99) rather than an inside switch.
- **Install motion-sensor-controlled spotlights** over the garage door and under the eaves. These not only discourage intruders, but also make it easier to carry in the groceries at night.
- **Place light posts or path lights along walkways**. Control bright lights with motion sensors and low-voltage lights with a timer or photocell so they stay on all night.
- **Make it difficult to extinguish lights.** Casual trespassers will usually be deterred by bright lights, but a professional thief will look for ways to shut off your security lights. Seven feet may be an attractive height for placing porch lights, but an intruder can easily reach that high. Place outdoor lights 9 or more feet above the ground.
- **Add standard-voltage light posts** to fortify your property. Low-voltage path lighting can be easily disconnected.
- **Install bright lights on motion-sensor switches** indoors behind a sliding glass door or large window. These will surprise intruders and alert you as well.

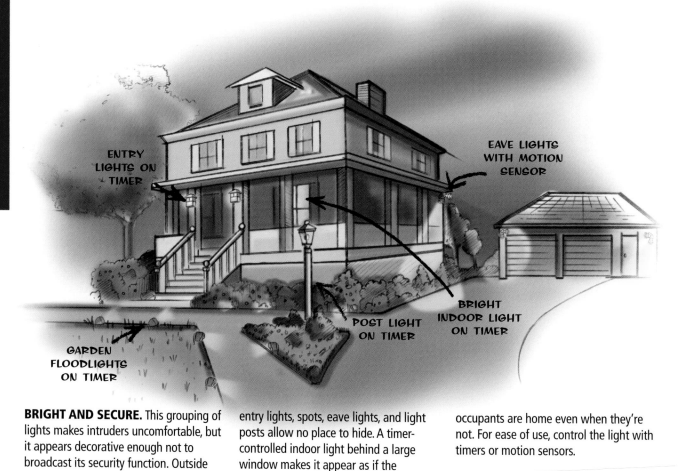

ENTRY LIGHTS ON TIMER

EAVE LIGHTS WITH MOTION SENSOR

POST LIGHT ON TIMER

BRIGHT INDOOR LIGHT ON TIMER

GARDEN FLOODLIGHTS ON TIMER

BRIGHT AND SECURE. This grouping of lights makes intruders uncomfortable, but it appears decorative enough not to broadcast its security function. Outside entry lights, spots, eave lights, and light posts allow no place to hide. A timer-controlled indoor light behind a large window makes it appear as if the occupants are home even when they're not. For ease of use, control the light with timers or motion sensors.

LIGHTING YOUR YARD

Lighting can emphasize your yard's best features. Begin by making a sketch of your property, including plantings, pathways, and outdoor structures. Spend an evening or two with a worklight and extension cord to try out some ideas. Vary the positioning. Outdoor lights may be suspended, mounted on poles, installed on the side of a deck or house, or placed under foliage. Consider these other options:

- **Try outdoor-rated rope lights.** Some rope lights (page 105) are designed for exterior use. Hang them loosely from post to post on a railing, stretch them taut along a fascia board, or spiral-wrap them—barber-shop style—around a pole or post.

- **Incorporate holiday lights.** Outdoor holiday lights—whether large and colorful or tiny white pinpricks—can be used year-round. Hang them high and fire them up for a party.

- **Use both standard-voltage and low-voltage lights.** Keep some standard-voltage lights around for times when you want to see clearly at night, but give yourself the option of using low-voltage lighting as well.

- **Experiment with color.** Use outdoor lenses and lightbulbs in various colors to set just the right mood. The results can be surprising, so take the time to experiment. Blue light resembles the cast of a full moon's light. Green light cast on a tree or shrub can give foliage a special luminescence. Reds, oranges, and yellows can evoke a warm, inviting feel.

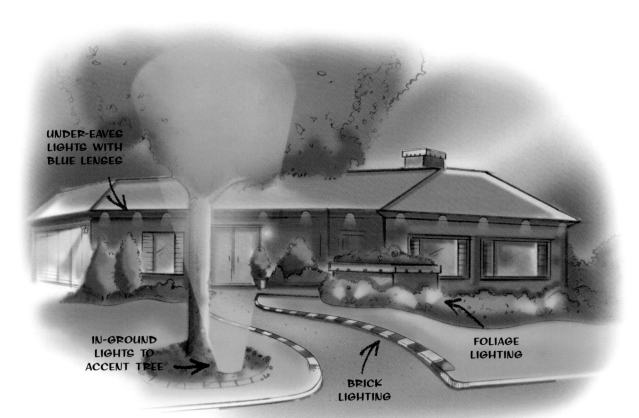

UNDER-EAVES LIGHTS WITH BLUE LENSES

IN-GROUND LIGHTS TO ACCENT TREE

BRICK LIGHTING

FOLIAGE LIGHTING

LIGHTS THAT EMPHASIZE FOLIAGE. Aim illumination toward attractive features of your yard, such as trees and plantings. Be sure these lights don't create an unpleasant glare for passersby. Bright under-eave lights are less harsh if blue lenses or bulbs are used. A tree can appear lit from within by an in-ground spotlight shining upward. Romanticize in-ground lights by simply dropping a few leaves on top of them. Lights that shine through flowering plants cast interesting shadows and highlight the colors of petals. Brick lighting defines the borders of a patio or driveway. All these elements can add to your home's security (opposite page) while they beautify your lot.

DECK AND PATIO LIGHTING

You'll find many light fixtures designed specifically for decks and patios at your home center. Position these lights to shine up from a patio surface, point down at a deck or stair treads, or sit atop posts and provide general illumination. The simplest way to add outdoor lighting is to plug in a string of low-voltage lights. However, keep in mind that these lights look temporary and are easily damaged.

120-VOLT FIXTURES

A flexible lighting system should combine low-voltage lights with standard 120-volt fixtures. Run standard-voltage cable or conduit in trenches (pages 164–165), hide it under decking or railing pieces, or drill long holes through posts and fish it through (page 166). Plan these installations to complement security lighting (page 94).

ROPE LIGHTS

Exterior-grade rope lights can be strung in fanciful patterns or in orderly straight lines. Unless you use a lot of them, they will be more decorative than bright. Plug them directly into a receptacle, or use an extension cord approved for outdoor use. Fasten them to wood posts and railings with galvanized fence staples.

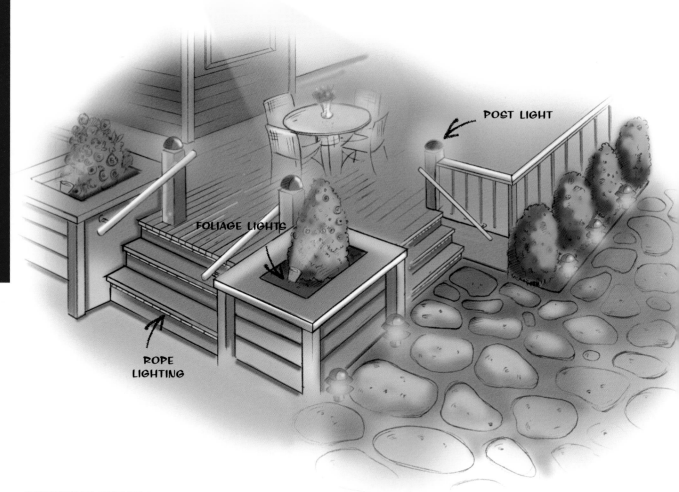

POST LIGHT

FOLIAGE LIGHTS

ROPE LIGHTING

LIGHTING TO SUIT EVERY MOOD AND PURPOSE. Treat outdoor diners to the same even lighting you would expect them to enjoy inside. Point several eave lights at the table, positioning them as high as possible. Rope lights are great as accents and to illuminate steps to help reduce the possibility of tripping. Post lights offer gentle highlights, while other lights give emphasis to specific features, such as plantings and flowers.

E lectrical components come in standard, interchangeable sizes. So, with little difficulty, you can remove an old fixture and replace it with one that has the style or features you want. Most of the fixture and device upgrades in this chapter can be done in less than a day. None of them calls for running new cable or changing electrical boxes. You won't have to calculate loads, make drawings, or tear apart your walls.

Typically, the only wiring you'll have to do involves detaching old wires and attaching new wires or leads. This is simple stuff, but don't be tempted to "hot-wire"— making connections without switching off the power. Always shut off the power and test for the presence of power before starting these upgrades.

EASY UPGRADES

CHAPTER SEVEN PROJECTS

A+ WORK SMARTER

SWITCH TO SMART LIGHTS

Upgrading standard wall switches to timers or motion sensors will help save energy by making sure that lights aren't burning when they don't have to be. Installation is easy and the results will save you money.

INSTALLING SPECIAL SWITCHES

SKILL SCALE

| EASY | MEDIUM | HARD |

SKILLS: Stripping and splicing wires, joining wires to terminals.

HOW LONG WILL IT TAKE?

PROJECT: Installing one of the special switches shown here.

EXPERIENCED . . . . . 15 MIN.

HANDY . . . . 30 MIN.

NOVICE . . . . 40 MIN.

STUFF YOU'LL NEED

TOOLS: Combination tool, lineman's pliers, screwdriver

MATERIALS: Specialty switch, wire nuts

Within an hour, you can install any of several clever switches that do everything from control a circuit at a preset time to switch on a light as you walk into a room. Instead of the familiar screw-down terminals on toggle switches, most special switches have **leads**—short lengths of stranded wire. Shut off power to the circuit. To splice leads, follow the directions on page 47.

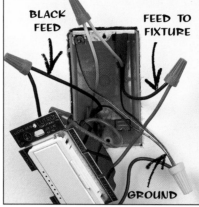

TOUCH-SENSITIVE DIMMER. With this switch, you dim or brighten a light by continuing to press the switch, rather than by turning a knob or operating a toggle. Connect the black lead to the black feed wire and the red lead to the black wire (or to the white wire painted black) running to the fixture. Connect the green lead to ground (page 13).

PILOT-LIGHT SWITCH. Use one of these for a garage light or an attic fan—anywhere you can't see the fixture while operating the switch. When the light glows, the fixture is on. Connect the black feed wire to the brass terminal where there is no connecting tab, and the other black wire to a brass terminal on the other side. Pigtail the neutral wires, and connect one to the silver terminal.

BUYER'S GUIDE

ANYWHERE SWITCH

This switch lets you control a fixture without having to run electrical cable. Wire the receiver inside the fixture, attach the sending switch "anywhere" on a wall, and put in a battery. You can even use these switches in a three-way setup (pages 148–150), and dimmers are also available.

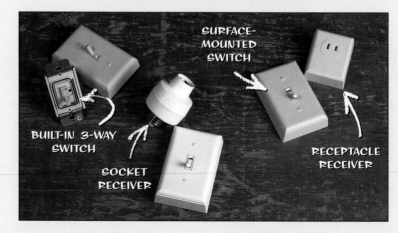

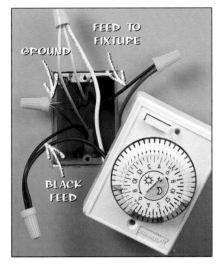

TIMER SWITCH. This switch turns outdoor lights on and off one or more times a day. Install one only if two cables (four wires, not counting the grounds) enter the box. (If only two wires exist, use a programmable switch instead.) Connect the black feed wire to the black lead and the other black wire (which goes to the fixture) to the red lead. Pigtail the whites, and connect them to the white lead.

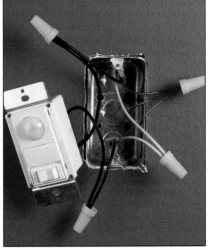

MOTION-SENSOR SWITCH. This switch turns on when its infrared beam senses movement. Adjust the time-delay feature to control how long the light will stay on. Wiring is the same as for a standard single-pole switch (pages 22–23), except that you connect to leads rather than to terminals.

CLOSE LOOK

ONE OR TWO?
If only one cable enters a switch box, then power runs to the fixture. Two wires—a black, and a white wire painted black—run from the fixture to the switch. If two cables enter the switch box, one brings power and the other runs to the fixture. With one cable, the black wire is the feed wire and the black-painted white wire leads to the fixture (page 20). Timer, combination, and pilot-light switches are among those that can be installed only if you have two cables in the box; there must be an unmarked white neutral wire.

COMBO SWITCHES

DOUBLE SWITCH. This device allows you to control two fixtures from a single switch box. Three cables enter the box: One brings power and the other two run to fixtures. Hook the feed wire to a terminal on the side that has a connecting tab. Hook the other two black wires to the terminals on the other side of the switch. Splice the white wires and the grounds.

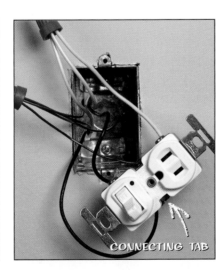

SWITCH/RECEPTACLE. This combines a switch and a grounded receptacle plug in a single switch box. The device usually is wired so the receptacle is hot all the time. Hook the feed wire to a terminal on the side with a connecting tab. Hook the other black wire to a brass terminal on the other side. Pigtail the white wires and hook to the silver terminal.

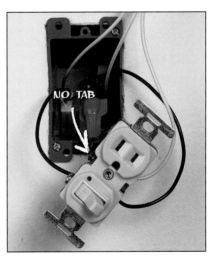

SWITCH/RECEPTACLE WITH RECEPTACLE CONTROLLED BY SWITCH. If you want the receptacle to turn off when the switch is off, reverse the positions of the black wires so the feed wire is on the side that does not have a connecting tab.

INSTALLING A GFCI BREAKER

EASY UPGRADES

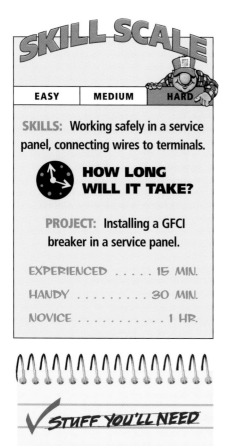

SKILL SCALE

EASY	MEDIUM	HARD

SKILLS: Working safely in a service panel, connecting wires to terminals.

HOW LONG WILL IT TAKE?

PROJECT: Installing a GFCI breaker in a service panel.

EXPERIENCED . . . . . 15 MIN.

HANDY . . . . 30 MIN.

NOVICE . . . . 1 HR.

STUFF YOU'LL NEED

TOOLS: Combination tool, linesman's pliers, screwdriver, flashlight

MATERIALS: GFCI breaker

The least expensive way to give a circuit GFCI protection is to install a GFCI receptacle. It can be wired to protect up to four additional receptacles (page 27). However, GFCI receptacles are notoriously short-lived. Because they will continue to provide power even when their ability to protect is lost, test them regularly.

For more reliable protection, install a GFCI circuit breaker. It's expensive, but it protects all the outlets on a circuit. Although you must feel comfortable about working on a service panel, installing a GFCI breaker can be easier than installing a GFCI receptacle. The latter is bulky and often requires a box extender.

See pages 14–15 for general safety instructions for working in a service panel. Always shut off power to the main breaker before you begin working.

SAFETY ALERT!

LIGHTNING PROTECTION

Lightning will seriously damage a roof, and if it hits the power line coming into your home, lightening may fry your service panel. Nothing will protect your home or electrical system from a direct hit, but some surge protectors will protect against nearby strikes. Or install an arrester system, which is basically an old-fashioned lightning rod. One or more rods are fastened to the highest points on the house, and a thick cable leads from them to the ground. The rod will conduct a reduced charge into the earth.

INSTALLING A GFCI BREAKER. Shut off the main breaker. (This de-energizes all the wires and circuitry after the main breaker, but the wires leading into the service panel will still be live.) Have a flashlight handy. Pull out the existing circuit breaker, loosen the terminal screw, and pull the wire out. Insert that wire into the GFCI breaker, and tighten the screw to clamp the wire tight. Detach the line's white wire from the neutral bus, and attach it to the breaker. Push the GFCI breaker into place as you would a standard breaker (page 80). Attach the curly white wire to the neutral bus bar, and restore power.

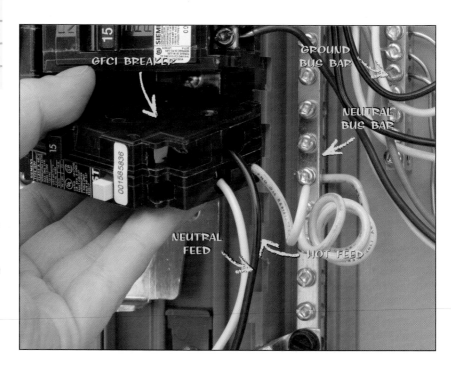

GFCI BREAKER

GROUND BUS BAR

NEUTRAL BUS BAR

NEUTRAL FEED

HOT FEED

ADDING SURGE PROTECTION

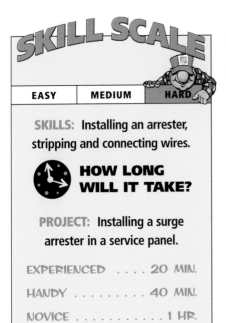

SKILL SCALE

EASY	MEDIUM	HARD

SKILLS: Installing an arrester, stripping and connecting wires.

HOW LONG WILL IT TAKE?

PROJECT: Installing a surge arrester in a service panel.

EXPERIENCED	. . . . 20 MIN.
HANDY	. . . . 40 MIN.
NOVICE	. . . . 1 HR.

✓ STUFF YOU'LL NEED

TOOLS: Combination tool, screwdriver, lineman's pliers, hammer.

MATERIALS: Surge arrester

O nce in a while, power supplied by your utility company may suddenly increase for a few milliseconds. This "surge" does not affect most electrical components, but it can damage sensitive electronic equipment, such as computers and televisions. A surge on your telephone line can destroy your modem and damage your computer. So buy surge protection. The higher a device's "joule" (a unit of strength of energy) rating, the better the protection. A surge arrester or protector will work only if the electrical system is grounded.

POWER STRIP SURGE SUPPRESSOR. An inexpensive device like this not only protects against surges but makes it easy to organize all those cords in a home office. Just plug it in. To protect a modem and computer, spend a little more for a device with a phone connection.

SURGE-PROTECTING CONSOLE. Available at stores that sell computers, under-monitor devices protect electronic equipment and phone lines from surges. They also help you organize the tangle of cords behind your desk.

A+ WORK SMARTER

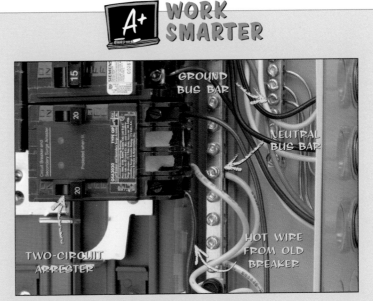

ARRESTER IN BREAKER BOX

To install an arrester that covers two circuits, shut off the main breaker. Remove two circuit breakers. Switch the breaker toggles to OFF. Push the breaker into place (page 80). Transfer the black wires from the old breakers to the terminals of the arrester. Attach the curly white wire to the neutral bus bar (make sure there are no kinks). Restore power and switch the arrester breakers on. Another type of surge arrester protects the entire panel.

EASY UPGRADES

101

INSTALLING TRACK LIGHTING

SKILL SCALE

EASY | MEDIUM | HARD

SKILLS: Laying out a track and anchoring with screws, stripping and splicing wires.

HOW LONG WILL IT TAKE?

PROJECT: Installing a track lighting system.

EXPERIENCED	2 HRS.
HANDY	4 HRS.
NOVICE	6 HRS.

✓ STUFF YOU'LL NEED

TOOLS: Combination tool, lineman's pliers, drill with screwdriver bit, tape measure, hacksaw

MATERIALS: Track system with lights, wire nuts, screws, plastic anchors

A track system is the most versatile of all ceiling fixtures. You can configure it in many ways (page 88), choose from several lamp styles, and position the lamps to suit your needs. To begin installation, remove the existing ceiling fixture to locate the track. If you don't have an existing ceiling fixture that is switched, see pages 146–147 for how to install a new one.

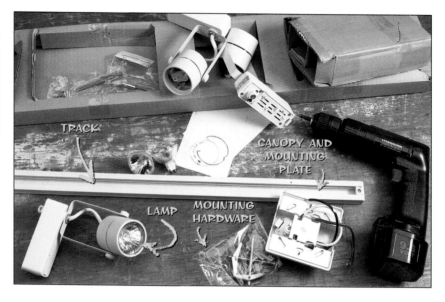

TRACK · CANOPY AND MOUNTING PLATE · LAMP · MOUNTING HARDWARE

PURCHASING A TRACK SYSTEM. Work with a salesperson; explain the size and configuration you want. Buy a kit that includes track, mounting plate, end cap, and canopy. You also may have to buy additional track and end caps, as well as L- or T- fittings for the corners. Choose the lights and bulbs when you buy the tracks. You can put different types of lamps on the same track, but be sure to purchase lamps made by the same manufacturer as the track—otherwise the two may be incompatible.

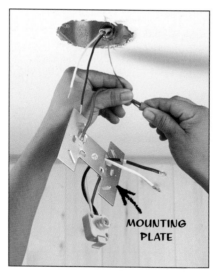

MOUNTING PLATE

1 INSTALL THE MOUNTING PLATE. Shut off power at the service panel. Use wire nuts to splice the house wires to the plate leads. Connect the ground wire to the plate and to the box if it is metal (pages 32–33). Push the wires into the box, and screw the plate to the box so it is snug against the ceiling.

2 MEASURE AND MARK FOR THE TRACK. At the mounting plate, measure to see how far the side of the track will be from the nearest wall. Mark the ceiling so the track will be consistently parallel to the wall. Use a framing square to draw lines if the track turns a corner.

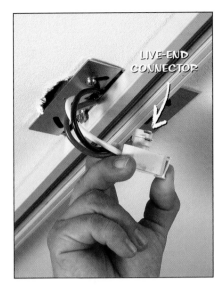

3 **ATTACH THE TRACK TO THE MOUNTING PLATE.** Have a helper hold the track in place against the ceiling and centered on the mounting plate. Drive the setscrews to anchor the track to the plate.

4 **SECURE THE TRACK.** Use a stud finder to locate joists. If the track is more than 4 feet long, have a helper hold one end while you work. Snap the track onto the plate, and drive a screw into every available joist. If there are no joists, drill holes every foot or so, insert plastic anchors, and drive screws into the anchors.

5 **TWIST ON THE LIVE-END CONNECTOR.** Insert the live-end connector and turn it 90 degrees until it snaps into place. Align the connector's two copper tabs with the two copper bars inside the track. Snap the plastic canopy over the track and mounting plate.

CLOSER LOOK

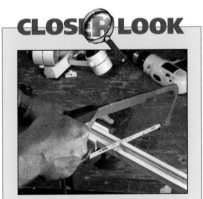

CUTTING TRACK

Tracks are available in standard lengths of 2, 4, 6, and 8 feet. If these sizes do not fit your needs, cut a track with a hacksaw or a sabersaw equipped with a fine-toothed metal-cutting blade. Clamp the track in a vise or hold it firmly with your hand as you cut. Cut slowly and take care not to bend the track while cutting. Reattach the plastic end piece.

6 **ATTACH A CORNER.** You can buy connectors to make 90-degree turns, T- shapes, or odd-angled turns. Slide the connector into the track that is already installed, slide the next track onto the connector, and attach that track to the ceiling. Cover all open track ends with end caps.

7 **TWIST ON A LIGHT.** This type of light twists into place in the same way as the live-end connector (Step 5). Another type has a metal arm that is twisted to tighten. Restore power, turn on the switch, and swivel the lights to position them for the best effect.

INSTALLING FLUORESCENT LIGHTING

SKILL SCALE

| EASY | MEDIUM | HARD |

SKILLS: Attaching with screws, stripping and splicing wires.

HOW LONG WILL IT TAKE?

PROJECT: Installing a fluorescent light fixture.

EXPERIENCED . . . . . 30 MIN.

HANDY . . . . 1 HR.

NOVICE . . . . 2 HRS.

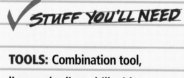

STUFF YOU'LL NEED

TOOLS: Combination tool, lineman's pliers, drill with screwdriver bit

MATERIALS: Fluorescent fixture, wire nuts, screws

INSTALLING A FLUORESCENT. Shut off power at the service panel. Remove the old fixture. Clamp the cable to a knockout in the new fixture and attach the fixture directly to the ceiling by driving screws into joists. Splice the fixture's wires to the incoming wires. Attach the cover plate.

BUYER'S GUIDE

CONSIDER FLUORESCENT LIGHTING OPTIONS.

■ If an old fluorescent light needs a new ballast (page 78), consider replacing the fixture. Newer fluorescents with electronic ballasts are trouble-free for decades.
■ Save energy costs by replacing an incandescent ceiling light with a fluorescent.
■ Fluorescent tubes offer a greater variety of light than ever (page 86). A diffusing lens further softens the light.

F luorescent lights often are installed without a ceiling box: Cable is clamped to the fixture, which substitutes for a box. However, some codes require that fluorescent lights be attached to ceiling boxes. Suspend the fixture or set it in a suspended ceiling grid and make the connections. Square or rectangular fixtures with long tubes are the most common. Other fluorescent fixtures are shaped like incandescents and use circular or U-shape tubes.

INSTALLING FLUORESCENTS IN A SUSPENDED CEILING. Fluorescent fixtures fit into the ceiling grid, taking up the space of a 2×2–foot (shown) or 2×4–foot ceiling tile. For smaller fixtures, install additional metal grid pieces and cut ceiling tiles to fit in either side. When you've established your power source for the lights (see pages 128–129 for how to extend the incoming line), install the grid, then attach the cable to the fixture, leaving more than enough cable to reach the power source. Connect to the power source before adding the tiles.

EASY UPGRADES

INSTALLING HALOGEN LIGHTING

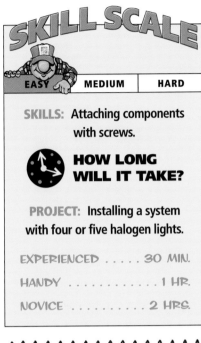

SKILL SCALE

EASY	MEDIUM	HARD

SKILLS: Attaching components with screws.

HOW LONG WILL IT TAKE?

PROJECT: Installing a system with four or five halogen lights.

EXPERIENCED . . . . . 30 MIN.

HANDY . . . . 1 HR.

NOVICE . . . . 2 HRS.

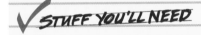

✓ STUFF YOU'LL NEED

TOOLS: Drill with screwdriver bit, combination tool, hammer

MATERIALS: Halogen kit (wire, lights, terminal block, transformer, cord switch), insulated staples

T o light counters or display shelves, consider a halogen puck light kit that plugs into a receptacle. A typical kit includes a transformer, cord, cord switch, several hockey-puck-shape lights that are attached to the underside of shelves or cabinets, and detailed instructions. If you don't like using a cord switch, plug the kit into a receptacle controlled by a switch (page 99), or alter a receptacle and run cable so one outlet can be switched off and on (page 143).

HALOGEN SAFETY TIPS

Halogens provide intense, almost glittering light and they get hot. Position them where people won't brush against them. Don't attach them to particleboard that may scorch. Use halogens in a closet only if you are sure they will always be 18 inches or more away from clothing or boxes. If you use them in small, enclosed spaces, such as shelves with glass doors, reduce the heat by replacing a 20-watt bulb with a 10-watt bulb. Drill ¼-inch air vent holes in the cabinet above the puck lights.

■ Halogens are very bright. Position them so they will be out of sight.

■ Never use a halogen without the lens, which filters UV rays.

■ Do not touch a bulb with your skin; natural oils will damage a bulb. Always handle halogen bulbs with a soft cloth.

A round-top stapler (page 180) makes it easier to attach the wires.

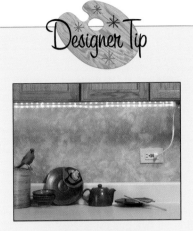

Designer Tip

INSTALLING HALOGEN ROPE LIGHTS

To install rope lights, simply staple a rope into place and plug it in. It's bright enough to use as under-cabinet countertop lighting and doesn't get as hot as "puck" lighting. Install rope lights in a straight line, or drape them in soft loops.

EASY UPGRADES

TERMINAL BLOCK

TRANSFORMER

INSTALLING PUCK LIGHTING. Drill ¼-inch holes to run wires, or plan to staple wires to the surface. Remove the covers from the lights, and mount the light bodies with screws. Snap on the trim rings, threading the wires through any holes. Attach a terminal block at a central location. Cut the transformer wires to length, strip the ends, and insert them into the terminal block. Plug the transformer into the wall. Staple down any loose wires.

INSTALLING TRAPEZE LIGHTS

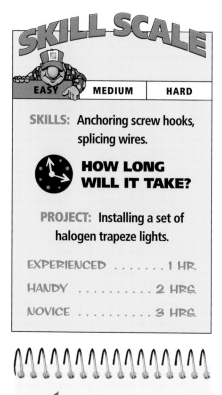

SKILL SCALE

EASY	MEDIUM	HARD

SKILLS: Anchoring screw hooks, splicing wires.

HOW LONG WILL IT TAKE?

PROJECT: Installing a set of halogen trapeze lights.

EXPERIENCED . . . . 1 HR.

HANDY . . . . 2 HRS.

NOVICE . . . . 3 HRS.

✓ STUFF YOU'LL NEED

TOOLS: Drill, screwdriver, longnose, side cutting pliers

MATERIALS: Trapeze light kit, wire nuts

These halogen fixtures are stylish, energy efficient, and—because you can easily point them to do the most good—versatile. The exposed wires are not dangerous because they carry very low voltage. Remove an existing ceiling light and attach a canopy transformer to the ceiling box (as shown in Step 2), or insert a plug-in transformer into a switched receptacle (page 143).

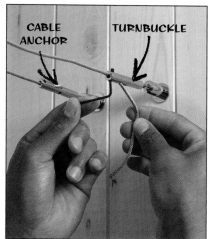

CABLE ANCHOR — TURNBUCKLE

1 **STRETCH THE CABLES.** Shut off power. Remove the light fixture (page 28) or install and run cable to a ceiling box (pages 122–132) where you plan to install the lights. Attach two cable anchors on the walls between which the unit will hang. Cut two lengths of cable to span the length of the installation. Fasten cables to the anchors and tighten the turnbuckle until the cables are taut.

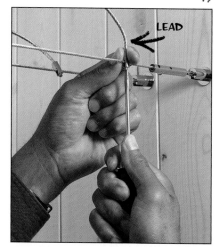

LEAD

3 **CONNECT THE LOW-VOLTAGE WIRES.** You may have to cut the low-voltage leads to the right length, restrip the clear insulation, and reattach the leads to the transformer. Clamp the leads onto the stretched cables using the fasteners provided with the kit. Attach the cover to the transformer.

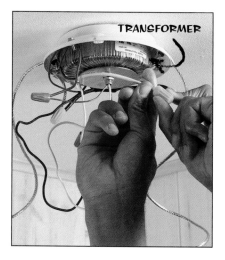

TRANSFORMER

2 **INSTALL THE TRANSFORMER.** Install a strap on the ceiling box (pages 28–29). Mount the transformer onto the strap. Splice the canopy transformer's red lead to the house's black wire, and splice white to white wires. Connect the green lead to the ground (page 13).

Go ahead and adjust the lights while the power is on. The voltage is so low you won't feel a thing.

4 **HANG THE LIGHTS.** Hold a halogen with a cloth (oil from your skin will damage it), and push the pins into a light arm. Slip the spring clamp over a wire, position the light arm on the cable, and clip the spring clamp onto the cable. Restore power and test.

GROUNDING RECEPTACLES

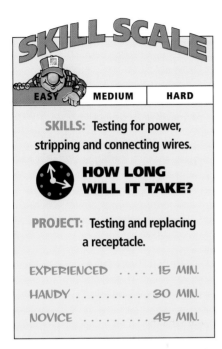

SKILL SCALE

EASY	MEDIUM	HARD

SKILLS: Testing for power, stripping and connecting wires.

HOW LONG WILL IT TAKE?

PROJECT: Testing and replacing a receptacle.

EXPERIENCED . . . . . 15 MIN.

HANDY . . . . 30 MIN.

NOVICE . . . . 45 MIN.

✓ STUFF YOU'LL NEED

TOOLS: Voltage tester or multitester, combination tool, screwdriver, longnose pliers

MATERIALS: Grounded receptacle, electrician's tape

I f a receptacle is ungrounded (with two slots only, and no grounding hole), its box may actually be grounded. If so, simply install a grounded receptacle. Do not install a grounded receptacle if the box is not grounded—you'll give the false impression the box is grounded when it is not. (For a description of grounding, see pages 11–13.)

If a box is ungrounded, ground it by running a #12 green insulated or bare copper wire to a cold-water pipe. Or, install a GFCI receptacle (page 27), which will provide greater protection than grounding.

1 **TEST AN UNGROUNDED RECEPTACLE FOR GROUND.** Scrape off any paint from the mounting screw. Insert one probe of a voltage tester or multitester into one receptacle slot, and touch the other to the mounting screw. Repeat test for other slot. If voltage is present, the box is grounded and you can install a three-hole receptacle.

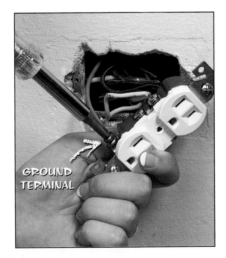

GROUND TERMINAL

3 **INSTALL A GROUNDED RECEPTACLE.** Snip off the stripped wire ends, which could break if they are bent again, and restrip. See page 20 for wiring directions. Be sure to test with a receptacle analyzer. If the test shows the receptacle is not polarized, switch the wires.

2 **TEST AGAIN.** If the first test is negative, remove the cover plate and repeat the first test, but touch the metal box, rather than the mounting screw, with one probe. If power is now indicated, you can install a grounded receptacle.

SAFETY ALERT!

AVOID GROUNDING ADAPTERS

This type of adapter works if it is connected to the mounting screw of a grounded box. But it's not much work to install a grounded receptacle, so there's no good reason to use an adapter. And, this installation is illegal in Canada.

EASY UPGRADES

107

INSTALLING MOTION-SENSOR LIGHTS

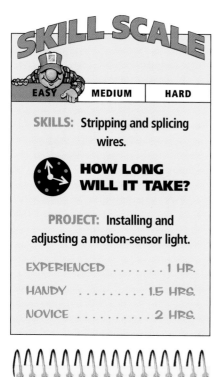

SKILL SCALE

EASY	MEDIUM	HARD

SKILLS: Stripping and splicing wires.

HOW LONG WILL IT TAKE?

PROJECT: Installing and adjusting a motion-sensor light.

EXPERIENCED . . . . 1 HR.

HANDY . . . . 1.5 HRS.

NOVICE . . . . 2 HRS.

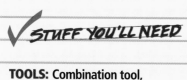

✓ STUFF YOU'LL NEED

TOOLS: Combination tool, screwdriver

MATERIALS: Motion sensor light, wire nuts, electrician's tape, perhaps a mounting strap

Motion-sensor lights greet you when you come home at night, and they discourage potential burglars. If you have an existing floodlight, they are easy to install. (To install an exterior box for a new light, see pages 162–163.)

Choose a fixture that lets you control the time and the sensitivity to motion. If the light is connected to a switch inside the house, you can override the motion sensor so the light stays on or off.

1 CONNECT THE LIGHT. Shut off power at the service panel, and remove the existing floodlight. If necessary, install a swivel strap (page 29). Run the wires through the rubber gasket, and splice them with wire nuts. While mounting the light to the box, position the gasket so it will keep the box dry.

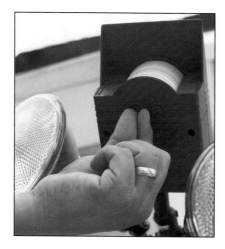

3 MAKE ADJUSTMENTS. To activate the motion sensor, manufacturer's instructions will probably tell you to turn off the wall switch, wait a few seconds, and turn it back on. Choose how long you want the light to stay on (ON TIME). There may be a control that keeps the light less bright for the amount of time you choose (DUAL BRIGHT). Set the RANGE to the middle position, and test how sensitive the motion sensor is by walking around near it. Adjust if necessary.

2 POSITION THE LIGHT. Restore power. Loosen the locknuts and twist the light until it is directed where you want it. Tighten the locknuts. At night, turn the light on permanently by flipping off the wall switch, then on again.

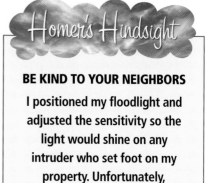

Homer's Hindsight

BE KIND TO YOUR NEIGHBORS

I positioned my floodlight and adjusted the sensitivity so the light would shine on any intruder who set foot on my property. Unfortunately, everyone walking on the sidewalk at night set off the light, making them feel like they were making a prison break! To keep peace with my neighbors, I readjusted the light.

8 PLANNING FOR NEW SERVICES

Once you are comfortable with projects like replacing devices and fixtures, you're ready to go—not boldly, but carefully—where few homeowners dare to tread. You're ready to add new electrical service. "New service" refers to running new cable. It can be as simple as tapping into a receptacle to add a new line (page 141) or as complex as installing several new circuits in a subpanel (pages 168–171).

This chapter helps you ask the right questions and come up with the best solutions so your new installation will do what you want it to do, and do it safely. You'll learn which tools and materials to buy, how to balance loads on a circuit and draw plans, and how to anticipate code requirements.

CHAPTER EIGHT TOPICS

BUYING TOOLS TO RUN NEW LINES

The money you pay for quality electrical tools will be minor compared to how much you'll save by doing the work yourself. To run new lines, you will need most of the tools shown here, as well as those on pages 42–43. Buy everything you need; the job will go easier.

FOR CUTTING INTO WALLS

These tools pave the way for installing electrical cable and boxes. The right tools, along with sharp bits and blades, will do the least damage to walls, saving you patching time afterwards. Consider buying a **corded power drill** with a $3/8$-inch chuck for large bits. (A cordless drill may not have the power or capacity to drill numerous holes in walls and framing.) Have several $5/8$-inch and $3/4$-inch **spade bits** on hand; they dull quickly. A **fishing bit** drills holes in hard-to-reach joists and studs. The **bender** helps you aim the bit where you want it to go. Once the hole is drilled, the bender has a **pulling attachment** that allows you to pull the cable with the bit.

Use a **drywall saw** to cut small holes in drywall. To cut through plaster and lath, use a **saber saw** with a fine-cutting blade, or use a **rotary-cutting tool**. A **flat pry bar** is ideal for trim removal and modest demolition. A **utility knife** is an essential all-purpose tool.

TIME SAVER

RENT A RIGHT-ANGLE DRILL

If you need to drill, say, 20 holes or more through studs or joists, you probably should rent a ½-inch right-angle drill.

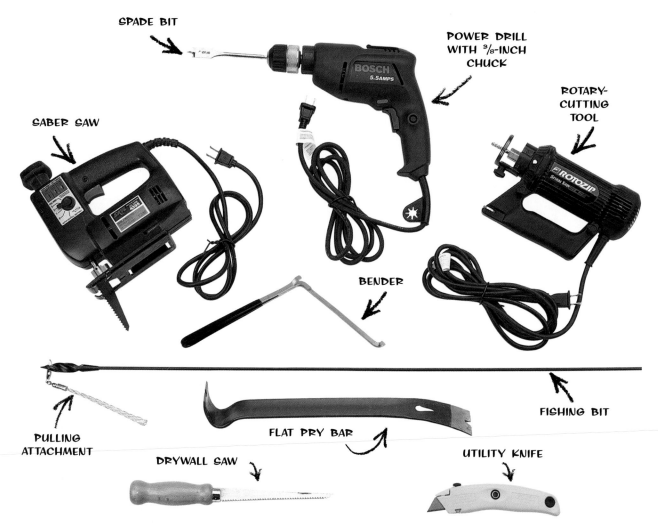

SPADE BIT

POWER DRILL WITH ⅜-INCH CHUCK

BOSCH 5.5AMPS

ROTARY-CUTTING TOOL

ROTOZIP

SABER SAW

SKIL

BENDER

PULLING ATTACHMENT

FLAT PRY BAR

FISHING BIT

DRYWALL SAW

UTILITY KNIFE

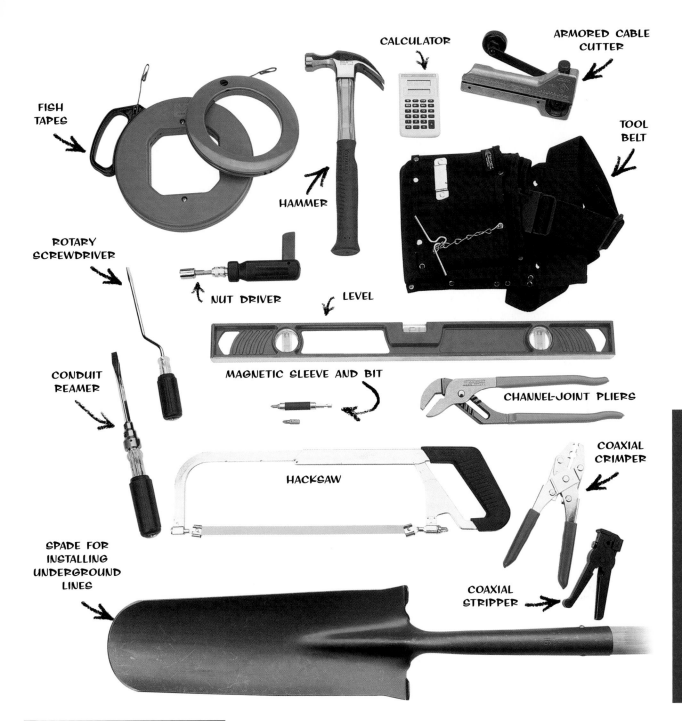

CALCULATOR

ARMORED CABLE
CUTTER

FISH
TAPES

TOOL
BELT

HAMMER

ROTARY
SCREWDRIVER

NUT DRIVER

LEVEL

CONDUIT
REAMER

MAGNETIC SLEEVE AND BIT

CHANNEL-JOINT PLIERS

COAXIAL
CRIMPER

HACKSAW

SPADE FOR
INSTALLING
UNDERGROUND
LINES

COAXIAL
STRIPPER

FOR RUNNING NEW LINES

Complete your kit with these relatively inexpensive tools. When working with armored cable, you may want to use an **armored cable cutter** (see page 124 for other options). For figuring circuit loads, use a small hand-held **calculator.** A **fish tape** helps you to run cable through finished walls and pull wires through conduit. Sometimes you need two tapes so that you can hook them together (page 131).

Buy an electrician's **tool belt** so you won't fumble around for tools. You'll use a **hammer** to tap locknuts tight onto cable clamps. If a box or fixture has bolts instead of screws, you'll need a **nut driver.** Use a **level** to mark cutouts on walls and square up boxes.

To drive screws quickly and firmly, nothing beats a drill with a **magnetic sleeve.** Insert small screwdriver bits into its tip, and they will be magnetized so you can drive screws with one hand. With a **rotary** screwdriver, you can drive or remove small screws on cover plates, switches, and receptacles in a flash.

You may need a pair of **channel-joint pliers** for handling connectors. Cut conduit with a **hacksaw** equipped with a professional-quality blade that will last longer. After cutting the conduit, remove burrs with a **conduit reamer.** For coaxial cable, use a **crimper** and a **stripper.** Use a **narrow spade** to excavate for outdoor cable.

COMMON CODE REQUIREMENTS

I t's your house and you're doing the work yourself. Why should a city inspector come around and tell you what to do? Codes and inspections are a sort of collective wisdom based on the experience of nearly a century of living with electricity. Those lessons have been incorporated into electrical codes. These codes exist to prevent house fires and injury from shocks, and to keep your electrical system running well.

MEETING NATIONAL, PROVINCIAL, AND LOCAL CODES

Whenever you run new electrical cable, your local building department will require you to get a permit and have the work approved by one of

its inspectors. Inspectors and building departments use the *National Electrical Code* (NEC) as the basis for most of their regulations. However, local standards often supplement or modify these basic rules.

You'll find some of the most common code requirements in the chart (opposite). The list is not complete, however, and you may need other sources of information.

You can buy a copy of the NEC, but it costs about $50 and is difficult to wade through. Most of its many pages deal with such items as commercial installations that homeowners will never encounter. Many useful handbooks on the NEC are available. Buy one, or borrow one from a library, that emphasizes residential installations.

COORDINATING THE TASKS

If you are building an addition to your house or gutting walls to remodel a kitchen or bathroom, you'll need to juggle carpentry, plumbing, and wall and floor finishing. Whether you do all or some of the work yourself, it's important that the various jobs are coordinated so that workers do not get in the way of each other and so that inspectors can see what they need to inspect. Aim for this sequence: **(1)** Install framing or gut the walls. **(2)** Run the rough plumbing, install electrical cable and boxes, and then call in the inspector. **(3)** Cover the walls with drywall, and paint. **(4)** Install the finish plumbing and electrical, and have it inspected.

PLANNING FOR NEW SERVICES

WORKING WITH INSPECTORS

Inspectors usually work with professional electricians who know codes and what is expected at inspections. Inspectors usually have a tight schedule and can't take time to educate you about what is needed. Their job is to inspect, not to help you plan your project. Take these steps to assure that the inspections go smoothly.

- Before scheduling an inspection, ask the building department for printed information about your type of electrical project. Make neat, readable, and complete drawings (pages 114–115), and provide a list of the materials.
- When you present your plans, accept criticisms and directives graciously. It usually does no good to argue—and the inspector does know more than you do. Make it clear that you want to do things the right way. Take notes while the

inspector talks to you so you can remember every detail of what needs to be done.

- Be clear on when the inspections will take place and exactly what needs to be done before each inspection. Before calling for an inspection, double-check that everything required is complete— don't make the inspector come back again. Don't cover up wiring that the inspector needs to see. If you install drywall before the inspection, you may have to rip it out and reinstall it after the inspection.
- Some building departments limit the kinds of work that a homeowner can do; you may have to hire a professional for at least part of a job. Others will let you take on advanced work only if you can pass an oral or written test.

WORKING WITH AN INSPECTOR

The first time I met with an inspector, the whole process bugged me.

I made a quick drawing and scrawled out a materials list. It was kind of a mess. The inspector got irritated, I argued back, and we were off to a bad start.

The fact is, he pointed out a dangerously overloaded circuit I was planning to install and a junction box I was going to drywall over. Both would have been dangerous. I came away grateful that he stood his ground.

CODES YOU MAY ENCOUNTER

Here's a quick summary of codes that are typical for household wiring projects. Follow them as you work up your plans and write your materials list.

These guidelines should satisfy most requirements, but keep in mind that your local codes might have different requirements. You probably will want to exceed requirements in order to provide your family with sufficient and safe electrical service.

The more you communicate your specific plans and techniques to your inspector, the less chance that you will have to tear out and do the job over again. It's better to be set straight by your inspector when the job is still on paper.

CABLE TYPE
Most locales allow NM (nonmetallic) cable for all installations where the cable runs inside walls or ceilings. Some areas require armored cable or conduit. If the cable will be exposed, many local codes require armored cable or conduit.

WIRE GAUGE
Use #14 wire for 15-amp circuits, and #12 wire for 20-amp circuits.

PLASTIC AND METAL BOXES
Many locales allow plastic boxes for receptacles, switches, and fixtures; but some require metal boxes. Boxes must be flush with the finished drywall, plaster, or paneling. Make sure boxes are large enough for their conductors (page 116).

RUNNING CABLE
NM and armored cable must be run through holes in the center of studs or joists so that a drywall or trim nail cannot reach it. Most codes require metal nail guards as well. Some inspectors want cable for receptacles to be run about 10 inches above the receptacles. NM cable should be stapled to a stud or joist within 8 inches of the box it enters. Once the cable is clamped to a box, at least ¼ inch of sheathing should be visible in the box, and at least 8 inches of wire should be available for connecting to the device or fixture.

CIRCUIT CAPACITY
Make sure usage does not exceed "safe capacity" (pages 62–63). Local codes may be stricter.

LIVING ROOM, DINING ROOM, FAMILY ROOM, AND BEDROOM SPECS
Space receptacles every 12 (4 meters) feet along each wall, and 6 feet (2 meters) from the first opening. If a small section of wall (between two doors, for example) is more than 3 feet (or 1 meter) wide, it should have a receptacle. For most purposes, use 15-amp receptacles. For convenience, rooms should have at least one light controlled by a wall switch near the entry door. However, pull chains for ceiling lights are acceptable. The switch may control an overhead light or one outlet of a receptacle, into which you can plug a lamp. Make sure the box you attach ceiling fans to can support the additional weight.

HALLWAY AND STAIRWAY SPECS
A stairway must have an overhead light controlled by three-way switches at the bottom and top of the stairs. If a hallway is more than 10 feet (3 meters) long, it must have at least one receptacle.

KITCHEN SPECS
Above countertops, space receptacles no more than 4 feet (1.22 meters) apart. Codes call for split-circuit receptacles above a countertop, which cannot be GFCI. Install one 15-amp circuit for lighting. Many codes require split receptacles on 15 amp circuits in kitchens and separate circuits for the dishwasher and refrigerator. A microwave must have a single 20-amp circuit.

BATHROOM SPECS
Any GFCI receptacle should be on its own circuit. Install the lights and fan on a separate 15- or 20-amp circuit.

GARAGE AND WORKSHOP SPECS
Install a 15-amp circuit for lights and a 20-amp circuit for tools. Install two 20-amp circuits if you have many power tools. Many areas require GFCIs in garages. Check your local code.

MAPPING A JOB

Building departments require detailed drawings and comprehensive lists of materials before issuing permits. To save yourself and the inspector aggravation, do your best to make your drawing clear and complete.

DRAW A PLAN

If you'll be wiring existing space, measure the rooms and make a scale drawing on graph paper. If you have blueprints for a new addition, use those. Make several copies of the floor plan so you can start over if you make mistakes. Include windows, doors, cabinets, and other obstructions.

Begin by drawing in all the switches, receptacles, and fixtures, using the symbols below. Then use color pencils to draw the cable runs. Use a different color for each circuit. Mark each cable—for example, "14/2 WG" for a cable with two #14 wires and a ground wire.

As you draw, make a list of materials, tallying the number of boxes, devices, and fixtures, and roughly figuring how much cable you will need.

DON'T FORGET

Check and double-check your drawing and your list.

- Make sure none of the circuits is overloaded (pages 116–117).
- See that switches are conveniently placed to easily turn on lights.
- Consider how each room will be used, and add devices where necessary. For instance, a home office with a computer should have a dedicated circuit.
- Make sure all your boxes will be large enough (page 116).
- Remember that if you add circuits, you may need to expand service with a subpanel (pages 170–171), a new service panel (page 172), or even a new line from the utility to your house. Determine how to run cable to the service panel or subpanel.

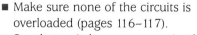

CLOSER LOOK

BASIC ELECTRICAL SYMBOL CHART.
Use these symbols as you plan your project.
They'll be easily understood by an inspector.

				Split receptacle
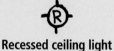	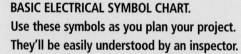		S_1	
Recessed ceiling light	**Duplex receptacle**	**240-volt polarized receptacle**	**Single-pole switch**	**Thermostat**
			S_2	
Ceiling light	**Split-wired duplex receptacle**	**Isolated ground receptacle**	**Double-pole switch**	**Indoor telephone**
			S_3	
Wall light	**GFCI receptacle**	**Weatherproof receptacle**	**3-way switch**	**Television jack**
			S_4	
Fluorescent ceiling light	**Switched receptacle**	**Service panel**	**4-way switch**	**Doorbell**

Fan	**Fourplex receptacle**	**Wall junction box**	**Switch with timer**	**Chime**

KITCHEN WIRING

12/2 14/2 14/3 14/2 14/2 14/3 14/2 14/3 14/3

15 AMP MICROWAVE

40 AMP 120/240 RANGE

15 AMP

S₁ S₁
S₃
S₃

40 AMP CIRCUIT

14/2

14/3

14/3

14/3

14/2

DISPOSAL

S S₃

DISH WASHER

14/2

CEILING FAN/LIGHT

14/3

REFRIGERATOR

14/2

14/3

PENDANT LIGHTS

14/2 14/2

S₃

14/2

MAKE A PLAN. Draw a floor plan of your project and make an extra copy or two. Use color pencils to distinguish your circuits. Add symbols for the various devices (they'll be useful when making your materials list). When you're satisfied with your plan, make a clean version to copy for the city and for your own use.

LOADING CIRCUITS CORRECTLY

When planning to add new service to your home, ask three important questions: First, will any individual circuit become overloaded as a result of adding service? Second, is your service panel large enough to accommodate any new circuits you will be adding? And finally, is the power entering your home from the utility company sufficient for your needs?

LOADING INDIVIDUAL CIRCUITS

If you will be extending an existing circuit to add a receptacle or a light fixture, make sure you won't overload that circuit. List all the receptacles, fixtures, and appliances on that circuit, and then add the wattage of the new service to determine whether you will be within "safe usage." (See pages 62–63 to make this calculation.)

SIZING UP A SERVICE PANEL

If existing circuits are not large enough to accommodate the new service you want to install, or if you will be wiring an addition, install new circuits.

Open your service panel (see page 60). If you see available blank slots, adding a circuit will be easy. If all the spaces are taken up, you may be able to add service by installing a tandem breaker (page 117).

Older homes with fuse boxes often receive 60-amp service from the utility company. If your wiring is less than 40 years old, your home probably has 100-amp service. Larger homes built in the last 15 years may have 200-amp service. The total amperage for your home is usually written on the main breaker or fuse.

GOT ENOUGH POWER?

Circuits can add up to more than their total rating. If you add up the amperage of all the breakers in the box, you will probably find that the total is more than the overall rating of the service panel. For example, a 100-amp service panel may have breakers totaling 220 amps. This does not mean that it is over capacity.

The amperage rating tells you how much amperage each hot **bus bar** (page 14–15) delivers. So each vertical row of breakers on a 100-amp box delivers 100 amps. And the breakers on a single bar can exceed

SIZING ELECTRICAL BOXES

Electrical codes calculate the cubic-inch capacity of boxes and then determine how many wires—and of what size—each box can accommodate.

A #14 wire occupies 2 cubic inches; a #12 wire occupies 2.25 cubic inches. (Canada: Use 1.5 cubic inches per #14 wire and 2 cubic inches per #12 wire.)

When counting the number of wires in a box, count a switch, receptacle, or any portion of a fixture that extends into the box as one wire.

As a general rule, buy large boxes unless you don't have room in your wall or ceiling. The bigger boxes don't cost much more, and they will give you room for upgrades in the future.

A METAL BOX used for a three-way switch holds nine #14 "wires"—two blacks, two whites, one red, three grounds, and one switch, for a total of 14½ cubic inches. Ground wires and wire nuts are not counted when calculating the size of the box.

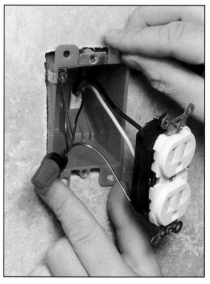

A PLASTIC BOX used for an end-of-the-line receptacle has only four #14 wires—a white, a black, and the receptacle, for a total of 8 cubic inches. Ground wires and wire nuts are not counted when calculating the size of the box.

the total capacity, because all the lights, fixtures, receptacles, and appliances will never run at the same time.

A home with more than 2,000 square feet will probably need more than 60-amp service. A house with less than 4,000 square feet that doesn't have electric heat or central air-conditioning probably needs no more than 100-amp service.

COMPUTING YOUR EXACT POWER NEEDS

To more accurately determine whether you have enough service, compute your home's electrical usage in watts. (Remember, watts = volts × amps; page 62). The *National Electric Code (NEC)* uses this formula.

■ Multiply the overall square footage of your home by 3 to determine lighting and receptacle needs.

■ For each kitchen appliance circuit and the laundry room circuit, add 1500 watts. Or add the total wattage used by all permanent appliances, such as the dishwasher, clothes dryer, and electric range.

■ Add it all up. Because you'll never use every outlet at full tilt at once, the NEC figures the first 10,000 watts at 100 percent and then adds 40 percent of the rest.

■ Add the wattage of either the central air-conditioning or the heating unit, whichever is greater.

■ Divide by 230 to figure how many amps you need. In the example below, 85 amps is needed, so 100-amp service is adequate.

If you need more power, consult with the utility company. They may need to install new wires, an expensive proposition.

PLANNING FOR NEW SERVICES

BUYER'S GUIDE

NEED MORE CIRCUITS?

Here are some options if you need more circuits than your service panel can provide.

ADD A TANDEM BREAKER

You may be able to install a tandem breaker, which supplies two circuits but uses only one slot. Check local codes to see whether tandem breakers are allowed for your service panel. You may need to install a subpanel or a new service panel instead (pages 170–172).

USE A BREAKER BOX EQUIPPED FOR NEW CIRCUITS

This box has plenty of room for new circuits. A standard 120-volt breaker will take up one slot, and a 240-volt breaker will use two spaces.

CLOSER LOOK

INFORMATIVE LABELING

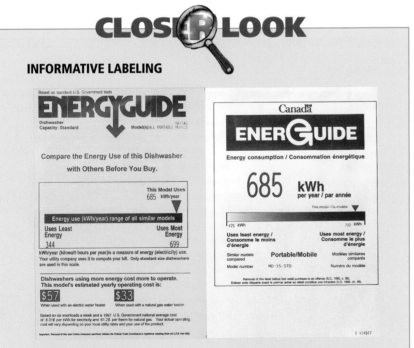

All major appliances such as air conditioners, water heaters, and refrigerators are required to carry labels that will tell you how energy efficient the unit is and how much it will cost to operate for a year under normal conditions and average use. Read the label carefully before you buy.

CHOOSING BOXES

All electrical connections must be contained inside a box. And all boxes—including junction boxes—must be accessible. Never cover a box with drywall or paneling. Some fixtures, such as recessed cans and fluorescent lights, contain their own boxes so connections can be made inside them.

Be sure to buy boxes large enough to avoid crowding the wires (page 116).

PLASTIC BOXES

In many areas, plastic boxes are the norm for all indoor residential wiring. They are inexpensive and quick to install. To install most new-work boxes, position and drive in the two nails. To install **remodel boxes** (boxes installed in walls already covered by drywall or plaster), see page 133–134.

Of course, you cannot ground a plastic box. For that reason, some local codes do not allow them, or they allow them only for certain purposes.

Some plastic boxes have holes with knockout tabs, so the cable is not held in place by the box. In that case, you must use cable clamps and staple the cable within 8 inches of the box. Other boxes have built-in metal or plastic cable clamps. Check local codes to see whether clamps are required.

Plastic boxes are easier to damage than metal boxes. When installing a new-work box, all it takes is one wrong swing with your hammer to crack the box. Never install a box that is cracked. Buy several extra boxes just in case.

Most plastic boxes are brittle, so don't use them where they are not built into the wall. The exception is an outdoor box made of especially strong PVC plastic.

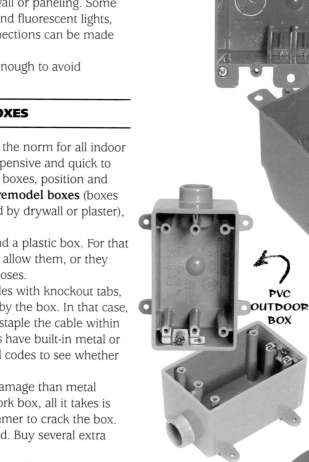

TWO-GANG BOX

THREE-GANG BOX

PVC OUTDOOR BOX

CEILING REMODEL BOX

ONE-GANG BOX

REMODEL BOX WITH EARS

METAL BOXES

Even if the local building department does not require metal boxes, you may prefer them because they are stronger and provide a better ground connection. Many codes require that all junction boxes—and all exposed boxes—be metal (with the exception of an outdoor PVC box, shown opposite). If a system uses conduit or armored cable and does not have a ground wire, the boxes must be metal in to provide a grounding path to the cable or conduit.

New-work metal boxes often have nailing brackets. Position the box, and drive screws or nails through the holes and into a stud or joist.

To open a knockout hole in a metal box, punch it with a hammer and screwdriver; then grab the slug from the inside with linesman's pliers and twist it off. Install a cable clamp if the box does not have built-in clamps.

Gangable boxes can be dismantled and ganged together to make space for two or more devices.

Install most switch boxes and ceiling boxes flush with the finished wall or ceiling surface. Install a junction box ½ inch behind the wall surface and add a **mud ring**—also called an adapter plate—which has screw holes for the cover plate. Choose a 4×4 junction box or a larger 4¹¹/₁₆-inch box. Make sure the cover plate or mud ring will fit the box. If a junction box holds only spliced wires and no device, cover it with a metal blank plate if it is exposed and a plastic blank plate if it will be enclosed in a wall.

Use round-cornered junction boxes called **handy boxes** if the box will be exposed on a basement or garage wall. Use metal cover plates. For a ceiling fan, use a fan-rated box (page 33).

1⁷/₈-INCH-DEEP HANDY BOX

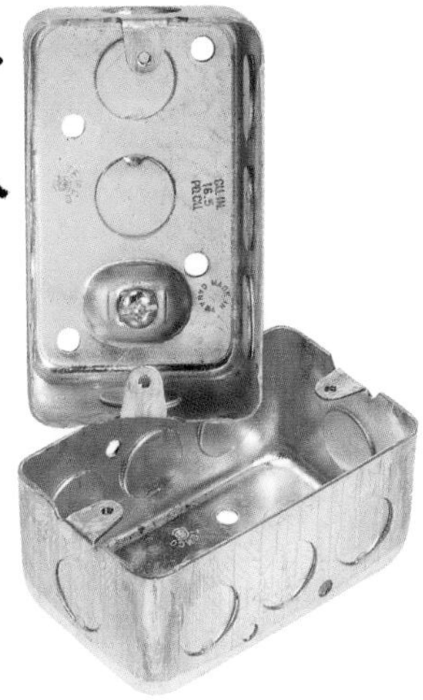

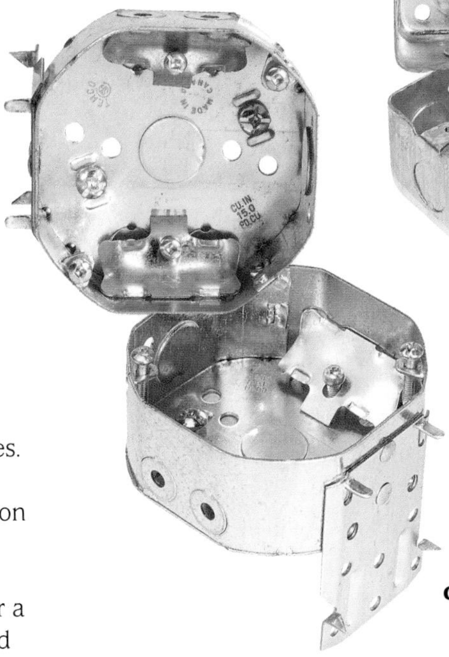

NEW-WORK OCTAGONAL BOX WITH BRACKET

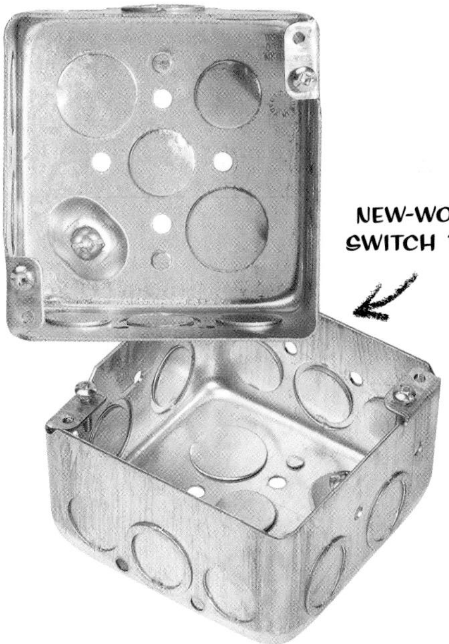

NEW-WORK SWITCH BOX

MUD RING

3½-INCH GANGABLE SWITCH BOX

INSTALLING STRAPS AND STAPLES

Cable, whether hidden in a wall or exposed, must be installed carefully to keep it from being damaged. Codes specify how and where each type of cable must be anchored. A **staple** holds the cable firmly without damaging its sheathing. Staples with plastic parts are better because they are less likely to damage the sheathing than the once-popular metal staples. Choose the right size staples to fit the cable.

When running cable along joists or studs, secure NM or armored cable at least 1¼ inches back from the front edge of the framing member (to protect it from drywall screws), using a plastic staple every 3 to 4 feet. Staple cable within 12 inches of a box that has a clamp and within 8 inches of a box that does not have a clamp. Never secure two cables with a single staple.

Use **drive straps** (below) for conduit; you can use **one- or two-hole straps** (below right) for either armored cable or conduit. When attaching a one-hole or a two-hole strap to wood, use a drill to drive in 1¼-inch screws. To anchor a strap to concrete, block, or brick, drill holes with a masonry bit and drive masonry screws into the holes.

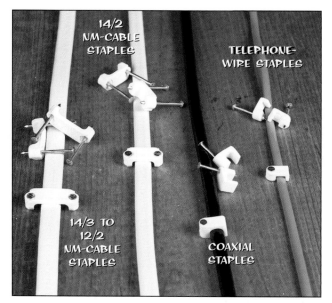

CHOOSING THE RIGHT STAPLE. A staple should hug the cable tightly without denting it. To attach it, position the staple over the cable, taking care that the nails do not touch the cable, and hammer in the nails. If you have a lot of telephone cable to install, purchase a staple gun that drives in round-topped staples (page 180).

ANCHORING CONDUIT TO WOOD. Use hook-like drive straps to attach conduit to wood. Hold the drive strap next to the conduit, and pound the strap in until it firmly grips the conduit. Place straps every 4 feet and within 12 inches of a box.

ANCHORING ARMORED CABLE TO WOOD. Fasten a strap in place with a 1¼-inch general-purpose screw. Use two-hole straps to install two parallel lines. Place the cable straps every 3 to 4 feet.

120

CHAPTER
9 *RUNNING NEW CABLE*

Before you start to run new lines, complete your wiring plan (Chapter 8) and get city approval for your project. Next, be sure to have a basic understanding of your home's electrical system (Chapter 1) and be comfortable with basic wiring techniques (Chapter 3). You'll find that installing boxes and cable in new framing is straightforward—even fun—once you've mastered the techniques. Running new lines in old walls is more challenging, especially if there isn't an attic or crawl space in which to run the lines. With planning, a few new skills, and the right tools, you'll get the job done right.

CHAPTER NINE PROJECTS

WORKING WITH NM CABLE

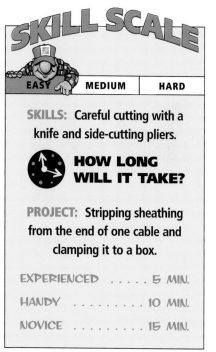

SKILL SCALE

EASY	MEDIUM	HARD

SKILLS: Careful cutting with a knife and side-cutting pliers.

HOW LONG WILL IT TAKE?

PROJECT: Stripping sheathing from the end of one cable and clamping it to a box.

EXPERIENCED . . . . . 5 MIN.

HANDY . . . . 10 MIN.

NOVICE . . . . 15 MIN.

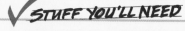

✓ STUFF YOU'LL NEED

TOOLS: Knife, lineman's pliers, side-cutting pliers

MATERIALS: Nonmetallic (NM) cable

You'll find nonmetallic (NM) cable easy to cut and quick to install. Just be careful when you remove the sheathing so you don't accidentally slit the wire insulation. If you do, cut off the damage and start again; otherwise you will get a short or a shock.

Whenever possible, strip sheathing before cutting the cable to length. That way, if you make a mistake you can try again.

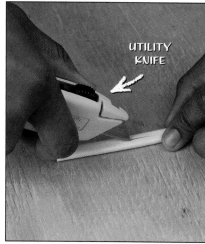

1 **SLIT THE CABLE WITH A KNIFE.** Flatten one end of the cable on a work surface. One side of the cable has a slight valley. With that side up, use a utility knife to start a cut about 3 inches from the cable end. Insert the blade into the middle of the valley (directly above the bare ground wire) so the blade just pierces through the sheathing. Slit to the end of the cable.

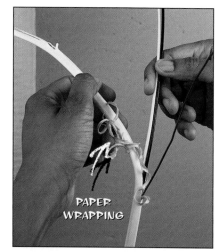

3 **REMOVE THE WRAPPING.** Pull the plastic sheathing back. Peel off any protective paper wrapping or thin strips of plastic, and cut them off.

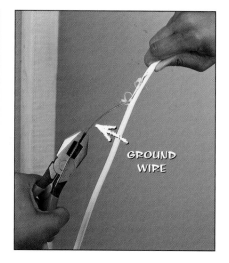

2 **PULL THE GROUND WIRE.** Cut or pull back the sheathing so you can grab the end of the ground wire with lineman's pliers. Hold the cable end in the other hand, and pull back the ground wire until you have made a slit in the sheathing about 12 inches long. This technique is common practice among electricians but pay extra attention because you can damage the ground wire while pulling it out.

TOOL TIP

USING A CABLE RIPPER

Use this tool to strip cable that is already installed in a box. Practice on scrap cable first to make sure the ripper doesn't cut too deeply and damage wire insulation.

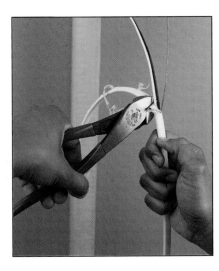

4 **SNIP THE SHEATHING.** Use side-cutting pliers, a combination tool, or the cutting portion of lineman's pliers to cut the sheathing.

5 **PULL THE CABLE INTO THE BOX.** Push the wires through the clip or clamp on the box (see right for the types you'll find). Pull the cable into the box so at least ¼ inch of sheathing shows inside.

CLOSER LOOK

FOUR WAYS TO ANCHOR NM CABLE TO A BOX

inspectors want to see 1/4 inch or more of sheathing in the box.

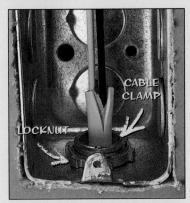

CABLE CLAMP. Buy clamps made for NM cable. Remove the knockout. Screw the clamp to the cable, then slip it through the hole and screw on the locknut. Tighten the locknut by tapping with a hammer and screwdriver. Or attach to the box first, slide the cable through the clamp, then tighten the screws.

POKE AND STAPLE. To run cable into many plastic boxes, you may need to push the cable past a plastic flap or knock out a plastic tab. Once you've inserted the cable into the box, staple the cable on a framing member within 8 inches of the box.

BUILT-IN CLAMP. Plastic boxes large enough to hold more than one device have internal clamps, as do most remodel boxes. Tighten the screw to firmly clamp the cable.

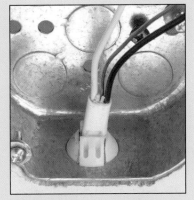

POP-IN PLASTIC CONNECTOR. Remove the knockout and push this connector in place. Then push the cable through and, if accessible, staple the cable within 8 inches of the box.

RUNNING NEW CABLE

WORKING WITH ARMORED CABLE

RUNNING NEW CABLE

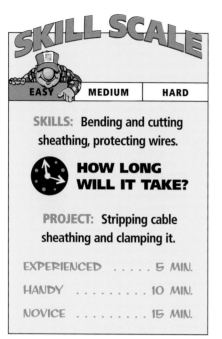

SKILL SCALE

| EASY | MEDIUM | HARD |

SKILLS: Bending and cutting sheathing, protecting wires.

HOW LONG WILL IT TAKE?

PROJECT: Stripping cable sheathing and clamping it.

EXPERIENCED . . . . . 5 MIN.

HANDY . . . . 10 MIN.

NOVICE . . . . 15 MIN.

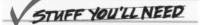

✓ STUFF YOU'LL NEED

TOOLS: Side-cutting pliers, screwdriver, hammer, channel-joint pliers, perhaps an armored cable cutter

MATERIALS: BX or MC cable, protective bushings

The features of flexible armored cable fall midway between NM and conduit. Armored cable is easier to install than conduit and less flexible than NM. It protects wires better than NM but won't turn away nails as well as conduit.

TYPES OF ARMORED CABLE

There are two types of armored cable. BX cable (also called AC90) is a type of armored cable with a ground wire. Older BX used heavy steel sheathing. Today's cable uses aluminum, which is lighter, is a better conductor, and is much easier to cut.

MC cable is like BX but with a green-insulated grounding wire. Some new building codes require using MC instead of BX for a sure ground.

WHERE TO USE IT

Some codes call for armored cable instead of NM. Others require NM or conduit where the cable is exposed. Run armored cable inside walls, and protect it from nails as you would NM cable. Armored cable will bend only so far, so use NM around wall corners (page 128) and around door jambs (page 131).

Caution: Sharp Edges! Handle armored cable carefully to avoid cuts.

1 **BEND AND SQUEEZE THE CABLE.** About 1 foot from the end, bend the cable and then squeeze the bend until the armor breaks apart slightly. If you have trouble doing this by hand, use a pair of channel-joint pliers.

TOOL TIP

NIPPING WITH A HACKSAW

Some people find stripping armored cable easier if they cut it with a hacksaw first. Barely slice through one of the ridges so you can see the wires but are sure you haven't nicked them.

USING AN ARMORED CABLE CUTTER

For large jobs, you may want to invest in this tool. Adjust the cutter for cable size, slip in the cable, and turn the handle to make a lengthwise cut.

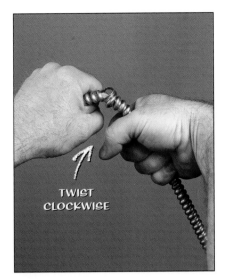

TWIST CLOCKWISE

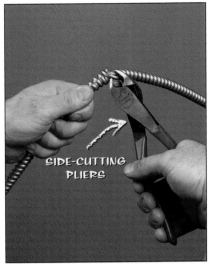

SIDE-CUTTING PLIERS

2 **TWIST THE CABLE.** Grasp the cable firmly on each side of the spot you want to cut. Twist the waste end clockwise until the armor comes apart far enough for you to slip in cutters. If you have trouble doing this with your bare hands, use two pliers.

3 **SNIP AND REMOVE THE ARMOR.** Cut through one rib of the armor with a pair of side-cutting pliers. Slide the waste armor off the wires. Keep your hands clear of sharp edges.

4 **TRIM SHARP ENDS.** Remove paper wrapping and plastic strips. Leave the thin metal bonding strip alone. Use side-cutting pliers to snip away pointed ends of sheathing that could nick wire insulation.

if bushings did not come with your cable, buy them separately.

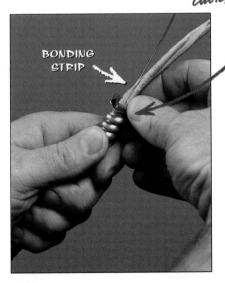

BONDING STRIP

ARMORED CABLE CLAMP

BONDING STRIP WRAPPED AROUND SHEATHING

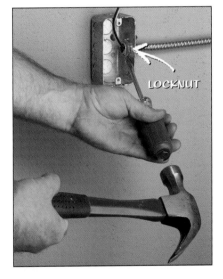

LOCKNUT

5 **SLIP ON THE BUSHING.** Slip a bushing over the wires. Slide it down into the armor so the bushing protects the wires from the sharp edges of the armor. If there is a bonding strip, ask your inspector what to do with it. Most inspectors want you to cut it to about 2 inches and wrap it over the bushing and around the armor, helping to ensure conductive contact between the armor and the box.

6 **ATTACH THE CLAMP.** Remove the locknut from an armored cable clamp. Slide the clamp down over the bushing as far as it will go, and tighten the screw. Double-check to make sure that none of the wires are in danger of being nicked by the armor.

7 **CONNECT TO THE BOX.** Remove a knockout slug from a metal box, and poke the connector into the hole. Slide the locknut over the wires, and tighten it onto the cable clamp. On BX cable, this connection is the ground—use a hammer and a screwdriver to tap the locknut tight.

RUNNING CONDUIT

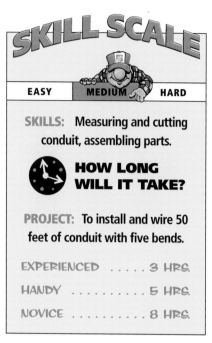

SKILL SCALE

EASY | **MEDIUM** | HARD

SKILLS: Measuring and cutting conduit, assembling parts.

HOW LONG WILL IT TAKE?

PROJECT: To install and wire 50 feet of conduit with five bends.

EXPERIENCED . . . . . 3 HRS.

HANDY . . . . 5 HRS.

NOVICE . . . . 8 HRS.

✓ STUFF YOU'LL NEED

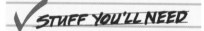

TOOLS: Conduit and fittings, wire, lubricant

MATERIALS: Screwdriver, lineman's pliers, hacksaw, conduit reamer, fish tape

C onduit is the most durable product for running wire. It's more expensive and time-consuming to install than cable, but it is no longer necessary to learn how to bend conduit. Ready-made parts make installation easier than ever. Use conduit on unfinished walls and ceilings where wiring will be exposed. Use electrical metallic tubing (EMT), or "thinwall" conduit, for most indoor installations and thicker intermediate metal conduit (IMC) for outdoor jobs. Plastic rigid nonmetallic conduit (PVC) is also used outdoors.

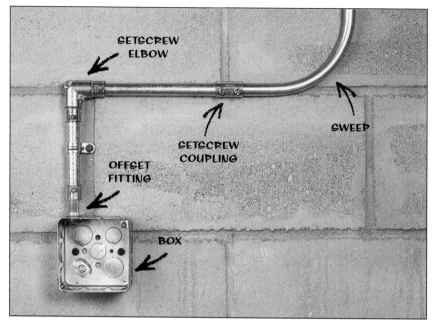

ASSEMBLING THE PARTS. Take a rough drawing of your installation to a home center or electrical supply store. Ask a salesperson to help you gather all the pieces you need. Generally, use ½-inch conduit for up to five #12 wires or six #14 wires, and ¾-inch conduit for more wires. (Larger conduit will make pulling easier, so consider buying ¾-inch in any case.)

Use **setscrew couplings and elbows** for indoor installations (you'll have to use compression fittings outdoors). If the conduit and the box are installed flush against a wall, you'll need an **offset fitting**. Use a **sweep** to turn most corners. At every four bends, provide access to the wires by installing a box or a pulling elbow (Step 5).

GREENFIELD. Also called flexible metal conduit, Greenfield is essentially armored cable without the wires. It is expensive, so use it sparingly in places where rigid conduit would be difficult to install.

PVC CONDUIT. In many areas, PVC is acceptable for indoor and outdoor installations. Cut it with a backsaw or hacksaw and a miter box. Glue the pieces together using PVC cement approved by an inspector.

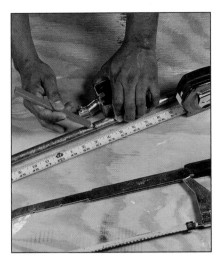

1 **MEASURE AND CUT.** Install the boxes first, then cut conduit to fit between them. At a corner, have a helper hold a sweep in place while you mark the conduit for cutting. Use a hacksaw with a fine-tooth blade to cut.

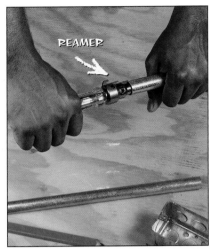

2 **REMOVE BURRS.** Ream out all burrs with a conduit reamer so the wires can slide smoothly past joints without damaging the sheathing.

3 **RUN FISH TAPE AND ATTACH THE WIRES.** Feed the fish tape through the conduit in the opposite direction from which you will pull the wires. Poke the wire ends through the fish tape's loop and bend them over in stair-step fashion. Wrap firmly and neatly with electrician's tape so the joint will not bind when it goes through a sweep.

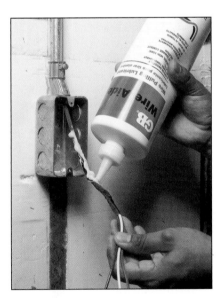

4 **SQUIRT LUBRICANT.** To make pulling easier on long runs, pour a bit of pulling lubricant on the wires. (Don't risk using substitute lubricants like dishwashing liquid or hand soap. Some can dangerously degrade wire insulation over time.)

5 **PULL THE WIRES.** Have someone feed the wires through one end while you pull the fish tape on the other end. Pull with steady pressure. Try to keep the wires moving, rather than starting and stopping. If you get stuck, back up a few inches to gain a running start.

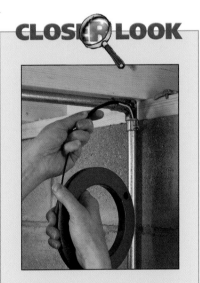

CLOSER LOOK

INSTALL PULLING ELBOWS

If the conduit will make more than three turns between boxes, install a pulling elbow to make fishing easier. Don't splice wires here; just use the opening to pull the wires through.

RUNNING NEW CABLE

127

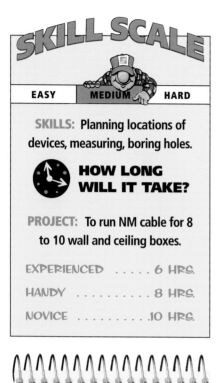

SKILL SCALE

EASY | MEDIUM | HARD

SKILLS: Planning locations of devices, measuring, boring holes.

HOW LONG WILL IT TAKE?

PROJECT: To run NM cable for 8 to 10 wall and ceiling boxes.

EXPERIENCED . . . . . 6 HRS.

HANDY . . . . 8 HRS.

NOVICE . . . . 10 HRS.

✓ STUFF YOU'LL NEED

TOOLS: Drill with ⅝-inch or ¾-inch bits, hammer, tape measure, level, longnose pliers, utility knife

MATERIALS: NM or armored cable, electrical boxes, protective nailing plates, cable staples, safety goggles

RUNNING NEW CABLE

R unning cable through bare framing members is far easier than fishing it through a wall finished with drywall or plaster and lath (pages 130–134). If the existing wall surface is flawed or if you also are installing plumbing in a remodeling job, it usually saves work to tear off all the drywall. Start anew, rather than living with a roomful of small wall patches.

INSTALLING CABLE THAT IS SAFE AND SECURE

Installing NM cable with plastic boxes is quick and easy—drill holes, run the cable through, and poke it into boxes. But don't run cable any old way. Safety concerns and codes dictate that it must be positioned out of harm's way, which means precise measuring and installing.

Choosing boxes and cable. Check local codes before buying materials. Codes may call for metal boxes, although plastic is fine in most areas. Assuming you will be installing ½-inch drywall after wiring, buy boxes that are easy to install ½ inch out from a stud or joist. Plan wiring carefully (Chapter 8), so you'll install the correct cables. For instance, use 14/2 for most general lighting and receptacles, 12-gauge for 20-amp circuits, and three-wire cable for three-way switches and split receptacles.

Placing holes. Local codes may specify the height at which cable for receptacles should be run, as well as where to put staples.

If an unfinished attic is above or a basement is below, run some of the cables there (page 132).

NAIL-ON BOX

1 INSTALL THE BOXES. Attach all the boxes before running cable. Receptacle boxes are usually positioned 12 inches above the floor and switch boxes 45 inches above the floor. (Many electricians set their hammer head down on the floor, using a hammer length to position floor-level receptacles.) Hold a nail-on box with its front edge positioned out from the stud the thickness of the drywall, and drive the two nails. Double-check to see that you've installed all the boxes. Walk around the room pretending to use all the switches.

SAFETY ALERT!

NEVER NOTCH

In a tight spot like this, you may be tempted to whip out the hammer and chisel and chop notches in the face of the studs so the cable runs easier. But the cable would then be dangerously exposed and severely bent at the corner. Instead, drill slightly larger holes, bend the cable before poking it in, and grab it with longnose pliers.

CORNER FRAMING

LONGNOSE PLIERS

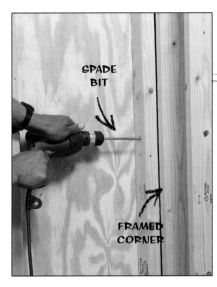

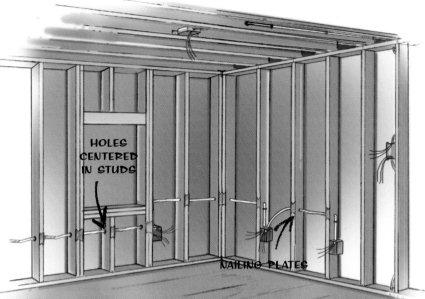

2 **DRILL THE HOLES.** Wherever possible, use a tape measure and level to mark studs and joists. Mark so holes will be in a straight horizontal line. Drill ⅝-inch holes for most NM cable and ¾-inch holes for three-wire cable or armored cable. A ⅜-inch drill works fine for small jobs, but give it a rest if it overheats.

A TYPICAL CABLE ROUGH-IN. Run cable in a straight horizontal line, 1 foot above the receptacles (areas under windows are an exception) or according to local code. To keep cable out of the reach of nails, drill all holes in the center of studs and at least 1¼ inches up from the bottom of joists. Nail on protective nailing plates for extra safety (they may be required for every hole). Even if you will only hang a light, install a ceiling fan box in case you choose to add a ceiling fan later.

3 **PULL THE CABLE.** To avoid kinks, keep the cable straight and untwisted as you work. When possible, pull the cable first and then cut it to length. If you must cut it first, allow plenty of extra length. Pull the cable fairly tight, but loose enough so there is an inch or so of play.

4 **PROTECT THE CABLE WITH NAILING PLATES.** These are inexpensive and quick to install. Be sure to nail one of these wherever the cable is within 1¼ inches of the front edge of the framing member. For added safety (and to satisfy some local codes), install nailing plates over every hole.

5 **STAPLE THE CABLE AND RUN IT INTO THE BOXES.** Staple cable tightly wherever it runs along a joist so it is out of the reach of nails. Staple within 8 inches of a plastic box and within 12 inches of a metal box. See pages 123 and 125 for clamping methods.

WIRING FINISHED ROOMS

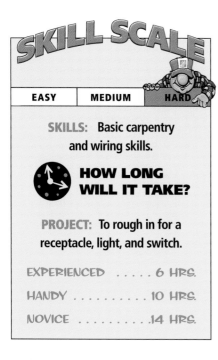

STUFF YOU'LL NEED

TOOLS: Electronic stud finder, drywall saw, saber saw, drill, hammer, screwdriver, fish tape, flat pry bar

MATERIALS: Cable, remodel boxes, safety goggles

You need the patience of a surgeon to run wiring through walls that are finished with drywall or plaster. At times, you'll feel like grabbing a hammer and knocking big holes in the wall to get at that darned cable. But remember that patching and painting walls are tedious and time-consuming tasks, so any steps you can take to minimize wall or ceiling damage will save you work in the long run.

FOLLOW THE EASIEST PATH

If you have an unfinished attic or a basement, run as much of the cable there as possible. If a basement or attic is finished, run armored cable instead of NM.

Use an electronic stud finder to locate joists and studs that may be in the way. You may be able to move a box a few inches to avoid an obstruction. Wherever possible, run cable parallel to studs or joists.

First, cut holes for the boxes (pages 133–134); then run the cable. Reach into the box holes with your hand, a fish tape, or a long drill bit in order to reach the cable.

If you plan to take power from an existing receptacle for your new service, make sure you will not overload the circuit.

1 DRILL A LOCATOR HOLE. Directly below a box from which you want to grab power, remove the base shoe and drill a ¼-inch hole through the floor. Poke a wire down through the hole.

2 DRILL UP THROUGH THE BOTTOM PLATE. Using the wire as a reference point, drill a 1-inch hole through the middle of the wall's bottom plate (a 2×4 lying flat on top of the flooring above).

3 HOOK THE CABLE. Open a knockout hole in the bottom of the box. Strip sheathing from the cable and attach a cable clamp (remove the locknut). Form the wires into a hook. Poke a fish tape or unbent coat hanger down through the knockout hole while a helper pushes the cable up. Hook and pull up.

RUNNING CABLE BEHIND A BASEBOARD. Use a flat prybar to remove baseboard molding. With a drywall saw, cut a channel in the drywall at least 1 inch shorter than the baseboard. Drill holes through the centers of the studs and run cable through the holes. Protect all holes with nail plates.

FISHING DRILL BIT

This tool eliminates the need for putting a hole in a wall or ceiling. Use the bit to drill a hole through the next stud or joist. While the tool is poking through the hole, hook a wire through the bit's hole, tape cable to the wire, and have a helper pull it back through the hole.

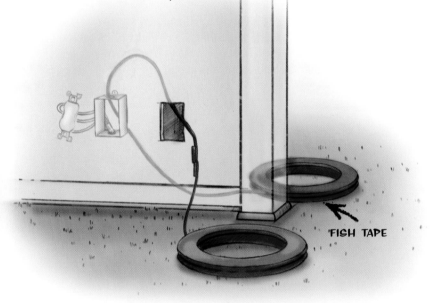

FISH TAPE

ADD PLATE HERE

RUNNING CABLE THROUGH A WALL. If the new box will be more or less directly behind the existing box from which you will grab power, you can avoid wall patching. Cut the hole for the remodel box. Remove the existing receptacle and punch out a knockout in the back or bottom of its box. Run one fish tape through the existing box and one through the new hole. Hook them together. Pull the tape back through the hole, and you're ready to pull cable from the hole to the box.

RUNNING CABLE AROUND A DOOR. If you have a slab floor and no access to the ceiling, this may be your only option. But check to see if this is OK with local codes. Remove casing from around a door and snake cable around. You may be able to slip the cable between the jamb and the stud. Or, drill a hole and run the cable in the cavity on the other side of the stud.

If the attic isn't used for storage, you may be allowed to lay cable on top of the joists if you install 1x4 strips on either side of the cable.

CABLE CLAMP WITHOUT LOCKNUT

FISH TAPE ATTACHED TO CABLE

RUNNING CABLE UP, OVER, AND DOWN. If the attic is accessible, drill a hole through the top plate. Run cable down through it to the hole for the new box directly below. Drill holes and run cable through the joists, over to the spot directly above the existing box from which you want to run power. Strip sheathing from the cable, install a cable clamp (without a locknut), and form the wires into a hook. Punch out a knockout hole and run a fish tape up the wall. Jiggle and slide the tape back and forth until it goes through the hole in the ceiling plate. Have a helper hook the cable to the tape and pull it into the box.

Homer's Hindsight

JUST CLAMP IT

When pulling cable through my finished wall and into the existing receptacle box, I didn't bother to attach a cable clamp. I figured, it's so much trouble just getting the cable through, why should I have to clamp it too? Well, the inspector disagreed, and I had to redo the job, installing the cable with a clamp. It's not that hard. Just pull the wires through, and the threaded part of the clamp will seat itself nicely in the hole.

WHERE THE CEILING AND WALL MEET. When there is no access from above or below, cut notches in the drywall and plaster, like this. Drill a 1-inch hole up through the center of the top plate. Bend the cable, poke it up through the hole, and grab it from the other side.

RUNNING NEW CABLE

INSTALLING REMODELING BOXES

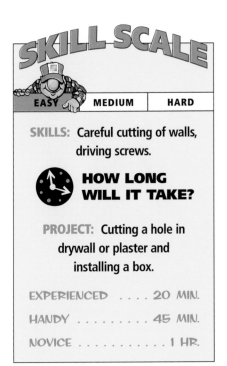

SKILL SCALE

EASY	MEDIUM	HARD

SKILLS: Careful cutting of walls, driving screws.

HOW LONG WILL IT TAKE?

PROJECT: Cutting a hole in drywall or plaster and installing a box.

EXPERIENCED . . . . 20 MIN.

HANDY . . . . 45 MIN.

NOVICE . . . . 1 HR.

✓ STUFF YOU'LL NEED

TOOLS: Electronic stud finder, utility knife, drywall saw, saber saw or rotary cutter, screwdriver, drill

MATERIALS: Remodeling (old-work) box, screws

When you run cable to install new devices in an old wall, you have several handy self-attaching boxes at your service. To use these **remodeling** boxes (also called **old-work** or **cut-in** boxes), you need only cut a hole, run the cable, clamp the cable to the box, and install the remodeling box.

To make sure you won't hit a stud or joist, before cutting a hole, drill a small bore in the wall, and probe with a piece of wire.

Cut the hole carefully, using one of the methods shown on this page. The hole will probably not be rectangular (page 141). The box should fit into the hole snugly, but not so tightly that you have to force it. If the hole is too wide, the box may not effectively attach to the drywall or plaster.

CUTTING A HOLE IN PLASTER WITH A SABER SAW. Cutting through a lath and plaster wall is difficult and often results in cracked plaster. Drill holes at each corner, and score the face of the plaster with a utility knife. Cut with a saber saw equipped with a fine-tooth blade. Press hard against the wall to reduce lath vibration.

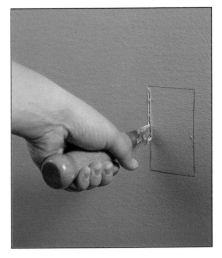

CUTTING A HOLE IN DRYWALL. Use a pencil to mark the location of the hole, and score the paper surface with a utility knife. Cut along the inside of the knife-cut with a drywall saw. The resulting hole will be free of ragged edges.

CUTTING A HOLE IN PLASTER WITH A SPIRAL CUTTING TOOL. Because of its rapidly rotating blade, this tool won't rattle your lath and loosen the plaster. To use this tool, set the base on the wall and tip the blade away from the surface while you let it come to full speed. Then tilt the blade gently into the wall. Have extra blades on hand; they dull quickly on plaster.

ATTACHING THE BOXES

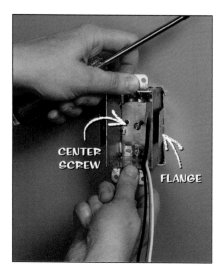

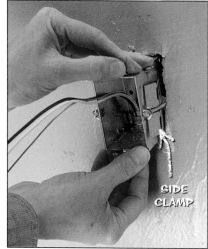

INSTALLING A BOX WITH SPRING FLANGES. If you buy this kind of box, make sure both flanges spring out firmly from the box. Push the box into the hole until the flanges are free to spring outward. As you tighten the center screw, the flanges should move toward you until they fit snugly against the back of the drywall or plaster.

INSTALLING A BOX WITH SIDE CLAMPS. After pushing the box into the hole, tighten the screw on each side. Each clamp extends behind the wall to hold the box in place.

USING MOUNTING BRACKETS. Push a metal box with plaster ears into the hole, then slip a bracket in on each side. Center each bracket behind the wall. Pull the bracket toward you until it's tight, push the box tightly against the wall, then fold the tabs into the box with your thumb. Tighten the tabs with pliers.

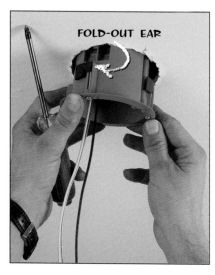

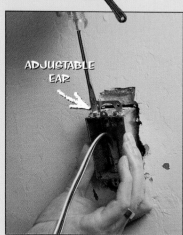

INSTALLING A BOX WITH FOLD-OUT EARS. These plastic remodeling boxes have ears that swing out behind the drywall or plaster. Push the box into the hole, then turn the screws clockwise until the ears clamp onto the back of the drywall or plaster. Switch boxes are also available with this same wall-grabbing mechanism.

1 ADJUST THE PLASTER EARS. Many metal boxes have adjustable ears. Cut the hole and chip out the plaster above and below so the ears will fit. Loosen the two screws and adjust each ear so the face of the box is flush with the wall surface. Tighten the screws.

2 ANCHOR THE BOX TO THE LATH. Lath cracks easily, so work carefully. Drill pilot holes, and drive short screws to anchor the ears to the lath. Expect to do some patching after using this method.

RUNNING NEW CABLE

PATCHING WALLS

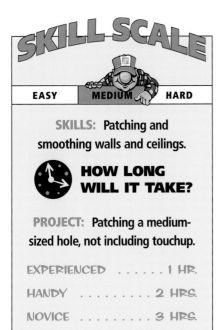
STUFF YOU'LL NEED

TOOLS: Putty knife, 4-inch and 8-inch taping blades, utility knife, sanding block

MATERIALS: Drywall, mesh patching tape, joint compound, spackling compound

The techniques shown on pages 130–132 help you minimize damage to walls, but patching drywall or plaster will probably be the finishing step in running cable.

Most homes built after the 1950s have walls covered with drywall—also called Sheetrock or wallboard. It's usually ½ inch thick and is fairly easy to patch. The time-consuming part is patching and smoothing the joint between the old and new surfaces.

An older home may have lath-and-plaster walls. The lath often splits or loosens and the plaster crumbles, making patching a challenge. Older homes may have a combination of the two: Old plaster walls are often covered with ¼- or ⅜-inch drywall.

HANDLING TEXTURED WALLS

Some drywall surfaces have a textured surface that is difficult to duplicate. You can cut out the pieces carefully and replace them with the original pieces when you are done wiring. You might get away with simply caulking the joints.

If a ceiling has a cottage-cheese-like appearance, a foam product has been blown onto it. You can buy a special patching compound to repair or recoat the ceiling, or hire a pro to recoat it.

1 **PATCH A SMALL HOLE IN DRYWALL.** Cut a new piece of drywall to fit the hole, or reuse the piece you cut out. If you do not have a stud or joist to screw to, cut a 1×4 about 4 inches longer than the hole. Place the piece behind the hole as shown. Drive 1¼-inch drywall screws to secure the patch.

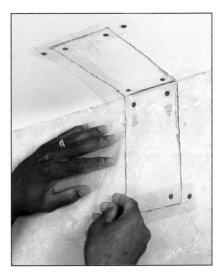

2 **TAPE THE JOINTS.** Cut pieces of fiberglass mesh patching tape and lay them over the joints. Apply joint compound and smooth it with a drywall taping knife. (Ready-mix joint compound is easy to use, but dry-mix compound is stronger and sets faster.)

3 **SAND THE PATCH.** Allow the compound to dry. Reapply the compound, feathering the edges. It will take several coats to smooth the joint. Sand the patch smooth with a drywall-sanding block. Prime and paint.

RUNNING NEW CABLE

135

RUNNING NEW CABLE

CLEAT

1 **CUT A HOLE FOR A DRYWALL PATCH.** Use a level or framing square to mark out a rectangle around the damage. Your marks should span from stud to stud or joist to joist. Cut with a drywall saw. Cut 2×2 or 2×4 cleats a few inches longer than the hole. Hold them against the back of the drywall as you drive 3-inch drywall screws into the framing.

2 **INSTALL THE PATCH.** Cut a patch to fit, about ¼ inch smaller than the hole in each direction. Attach the patch with 1¼-inch drywall screws. Cover the joints with tape, apply joint compound (right), and sand (page 135).

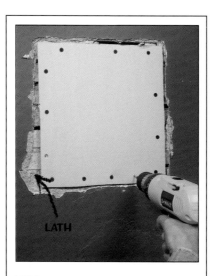

LATH

1 **PATCH PLASTER WITH DRYWALL.** Remove loose plaster. Tap with a hammer to excavate a rough rectangular shape. If the lath is solid, don't expose studs or joists. For the patch, use drywall that is the same thickness as the plaster. Cut the patch roughly to size, and attach it to the lath with 1¼-inch drywall screws.

2 **FILL THE GAP WITH JOINT COMPOUND.** You can mix it with perlited gypsum (see "Cutting a Channel in a Plaster Wall," above). Apply mesh tape to the joints, then apply joint compound. For best results, apply several feathering coats of joint compound, scraping and sanding between coats.

WORK SMARTER

CUTTING A CHANNEL IN A PLASTER WALL

You might cut a narrow channel through a plaster wall to slip a cable through. To fill the gap, combine dry-mix joint compound ("90" or "45") with an equal amount of perlited gypsum. Mix the two with water, and you'll have a paste that won't sag when you apply it. Force the paste into the cavity with a putty knife. Allow it to dry, then apply subsequent coats of joint compound.

Homer's Hindsight

BLENDING A PATCH

The old plaster walls in my bedroom were rough, but they had a fairly uniform texture. They looked fine until the day I patched up one area. I did such a smooth job that the patch emphasized the flaws on the rest of the walls. I had to go over the patch with a looser, rougher coat.

INSTALLING A JUNCTION BOX

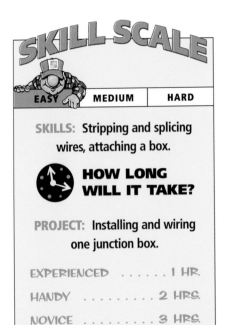

SKILL SCALE

EASY	MEDIUM	HARD

SKILLS: Stripping and splicing wires, attaching a box.

HOW LONG WILL IT TAKE?

PROJECT: Installing and wiring one junction box.

EXPERIENCED ...... 1 HR.

HANDY .... 2 HRS.

NOVICE .... 3 HRS.

✔ STUFF YOU'LL NEED

TOOLS: Combination tool, lineman's pliers, screwdriver, drill, voltage tester

MATERIALS: Junction box with cover, wire nuts, screws

Install a junction box wherever wires must be spliced. Keep the box accessible—never bury it in a wall or ceiling. Junction boxes are usually flush-mounted to walls or attached to attic, basement, or crawl-space framing. But you can set one inside a wall as you would a switch box. Cover the junction box with a blank plastic cover plate.

1 ATTACH THE BOX. Shut off power to the wires that you will be splicing. Anchor the box with screws. To attach the box to a masonry surface, drill holes with a masonry bit. Drive masonry screws.

GROUNDING PIGTAIL

2 WIRE THE BOX. Strip cable sheathing and clamp the cable, or connect conduit. Strip wires and connect them with wire nuts. If the box is metal, make a grounding pigtail and connect it to the green grounding screw.

COVER PLATE

3 COVER THE BOX. Fold the wires into the box and attach the cover plate. To do so, loosen the screws at two corners of the box, hook the cover plate on first one screw and then the other, and tighten the screws.

A+ WORK SMARTER

USE A METAL COVER PLATE IN UTILITY AREAS

If a receptacle or switch is in an exposed box, use a metal rather than a plastic cover plate. You may need to break off the device's metal "ears." Attach the devices to the cover plate first, and then attach the cover plate to the box.

137

INSTALLING RACEWAY WIRING

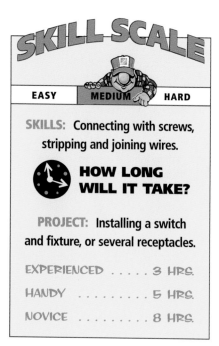

SKILL SCALE

EASY   MEDIUM   HARD

SKILLS: Connecting with screws, stripping and joining wires.

HOW LONG WILL IT TAKE?

PROJECT: Installing a switch and fixture, or several receptacles.

EXPERIENCED . . . . . 3 HRS.

HANDY . . . . 5 HRS.

NOVICE . . . . 8 HRS.

✔ STUFF YOU'LL NEED

TOOLS: Drill, screwdriver, hacksaw, combination tool, longnose pliers

MATERIALS: Box extender with cover plate, channel, fittings, fixture bases, fixture box, wire, clips, new devices, plastic anchors

R aceway wiring is an easy way to install a switch, fixture, or receptacle when cutting into a wall is difficult or appearances aren't important. It will spare you the hassle of cutting into walls, drilling holes, fishing cable, and patching the walls.

GATHERING THE PARTS

Take a drawing of your proposed installation to a home center or electrical supply source, and ask a salesperson to help you assemble all the parts. Choose metal, which is paintable, or plastic, which is not.

You'll need a starter box for each device, channel, L and T connectors, receptacle or switch boxes, and perhaps a fixture box. Buy plenty of wire. Use green-insulated wire for the ground, never bare copper.

STARTER PLATE

EXISTING RECEPTACLE

1 **INSTALL THE STARTER BOX.** Shut off power to the circuit. Pull out a receptacle, and mount a starter plate on the wall behind it. Install new raceway boxes for receptacles, switches, and fixtures in the same way.

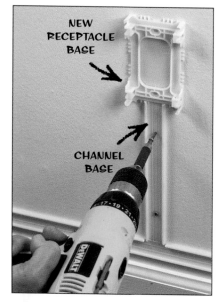

NEW RECEPTACLE BASE

CHANNEL BASE

2 **CUT AND ATTACH THE CHANNEL BASE.** Use a hacksaw to cut pieces of channel to fit between the boxes. Attach the channel base to the wall with screws driven into studs or plastic anchors. Use fittings at all corners.

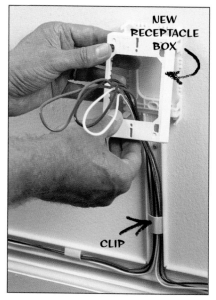

NEW RECEPTACLE BOX

CLIP

3 **RUN THE WIRES.** Place wires in the channel base and secure them with clips about every foot. Leave 8 inches of wire at each box to make connections. Snap any device boxes onto the bases and fasten them with screws driven into studs or plastic anchors.

CHANNEL COVER

4 **WIRE THE DEVICES AND FIXTURES.** Snap the covers onto the channel base and the corner pieces. Strip the wire ends and connect them to the terminals just as you would for standard wiring. Install cover plates, restore power, and test.

FISHING WIRES IN METAL RACEWAY. Metal channels do not come apart in two pieces. Install clips on the wall, and snap the channel into the clips. Fish wires through the channel. If you can't shove the wires through, you might have to use a fish tape.

ADDING A LIGHT AND WALL SWITCH USING RACEWAY WIRING. To add a switched ceiling light to a room, you need a nearby receptacle. Install a ceiling fixture base, making sure that it is firmly attached to joists in the ceiling. Install the raceway switch base, and run a channel from the receptacle to the switch base and on to the fixture base. Run wiring and add the boxes. Make wiring connections as shown on page 142, and install the devices.

Use a similar arrangement to add a wall switch to a pull-chain light fixture. Install a raceway switch box at a convenient height. Remove the ceiling fixture, and install a raceway fixture box onto the ceiling box. Run the channel and two black wires from the switch to the fixture, making connections as shown on page 146.

BUYER'S GUIDE

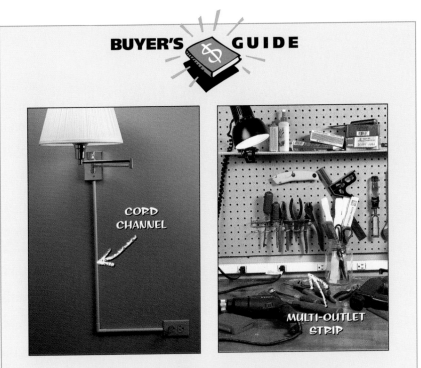

CORD CHANNEL

MULTI-OUTLET STRIP

TRY THESE QUICK AND EASY PRODUCT OPTIONS
Cord channel (above left) encases and protects lamp cord that must be run along a wall. A multi-outlet strip (above right) is a sort of super extension cord, with a grounded receptacle every foot or so. No wiring is required to install these products. The channels mount to clips, or they stick to the wall with tape backing.

INSTALLING NEW SERVICES

ew construction projects are more satisfying than adding electrical devices or fixtures to your home. Most of these jobs take less than a day, yet make big improvements in your family's quality of life. The installations in this chapter rely on the skills and knowledge taught in the first two-thirds of the book. Refer to earlier chapters for specific instructions on the projects that follow.

Whenever adding new services, follow these important guidelines:

- Turn off power to the circuit you are working on, and test all open boxes to make sure no is power present.

- Be sure the new service will not overload your circuit.

- Follow local codes for running cable and installing boxes. Obtain a permit from your building department every time you install new cable.

CHAPTER TEN PROJECTS

INSTALLING A NEW RECEPTACLE

SKILL SCALE

EASY	MEDIUM	HARD

SKILLS: Running cable through walls, stripping and splicing wires, attaching boxes.

HOW LONG WILL IT TAKE?

PROJECT: Installing a new receptacle in a finished wall.

EXPERIENCED . . . . . 2 HRS.

HANDY . . . . 3 HRS.

NOVICE . . . . 5 HRS.

STUFF YOU'LL NEED

TOOLS: Drill, drywall saw or sabersaw, screwdriver, lineman's pliers, combination strippers

MATERIALS: Cable and clamps, remodel box, staples, receptacle, wire nuts, electrician's tape

The easiest way to install a new receptacle is to tap an existing receptacle for power, as shown here. Before you do this, make sure you will not overload the existing receptacle's circuit (page 116–117).

If you can't pull power from a nearby receptacle, you may be able to tap into a junction box above or below the room. You also can pull power from a light fixture or switch—whichever has power entering its box.

As a last resort, you may have to run cable all the way back to the service panel and install a new circuit breaker (pages 168–169).

As with many electrical projects, patching and painting walls afterward can be more trouble than the wiring. See pages 130–132 for tips on reducing damage to your walls.

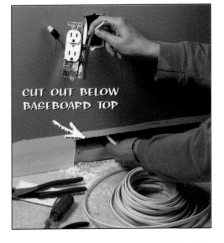

CUT OUT BELOW BASEBOARD TOP

1 PLAN A PATH FOR THE CABLE. Choose a path that will cause minimal damage to the walls, such as running cable behind a baseboard (shown). Remove the baseboard and cut away the drywall. Shut off power to the circuit. Remove a knockout in the receptacle's box from where you'll take power, add a connector, then fish the cable.

REMODEL BOX

2 RUN THE CABLE THROUGH THE HOLES. Cut a hole for the new receptacle box. Drill holes in the centers of studs for the cable to pass through. Strip 6 to 8 inches of sheathing from either end of the cable. Punch out a knockout hole and clamp the cable to the existing box. Run the cable into a remodel box and attach the box to the wall (pages 130–134). Either clamp the cable to the box or staple the cable near the box.

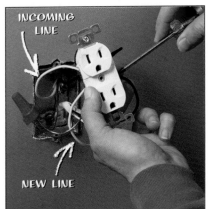

INCOMING LINE

NEW LINE

3 WIRE THE RECEPTACLES. If the existing receptacle is at the end of the run (shown), attach the black wire to the brass terminal and the white wire to the silver terminal. If the receptacle is in the middle of the run, no terminals will be available; use pigtails to connect to power (page 49). Wire the new receptacle—white to silver, black to brass. Connect the grounds (pages 12–13). Restore power, then test.

INSTALLING NEW SERVICES

ADDING A WALL SWITCH

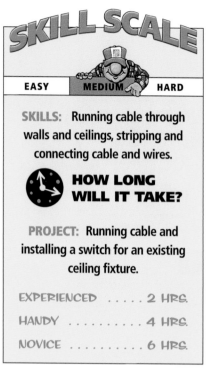

SKILL SCALE

EASY	MEDIUM	HARD

SKILLS: Running cable through walls and ceilings, stripping and connecting cable and wires.

HOW LONG WILL IT TAKE?

PROJECT: Running cable and installing a switch for an existing ceiling fixture.

EXPERIENCED	..... 2 HRS.
HANDY	.... 4 HRS.
NOVICE	.... 6 HRS.

STUFF YOU'LL NEED

TOOLS: Drill, drywall saw or saber saw, fish tape, screwdriver, lineman's pliers, strippers

MATERIALS: Cable and clamps, remodel box, staples, receptacle, wire nuts, electrician's tape, cover plates

Wiring a wall switch to a pull-chain ceiling fixture is simple; the challenge is running cable from the fixture to the switch. You will probably need to cut a hole near the fixture so that you can reach behind its box and clamp cable to it. Consider covering the hole with a medallion (page 29).

PULL CABLE HERE...

...THEN HERE

1 RUN CABLE. Shut off power to the circuit supplying the fixture. Plan a cable pathway that crosses as few studs or joists as possible. You may have to cut access holes to run cable through framing (page 132).

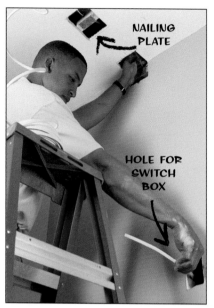

NAILING PLATE

HOLE FOR SWITCH BOX

2 RUN CABLE TO THE SWITCH BOX. Add nailing plates where you bore holes in framing. Cut a hole for a remodel switch box and pull the cable through. Strip the wires.

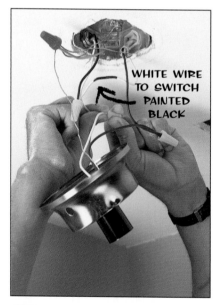

WHITE WIRE TO SWITCH PAINTED BLACK

3 WIRE THE FIXTURE. First, connect the ground (pages 12–13). Remove the old black wire from the fixture lead, and splice it to the new white wire and mark it black. Splice the new black wire to the fixture's black lead.

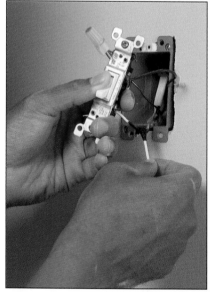

4 CONNECT THE GROUND AT THE SWITCH. Attach both wires to the terminals and mark the white wire black. Restore power to the circuit, and test.

CONTROLLING A SINGLE OUTLET WITH A SWITCH

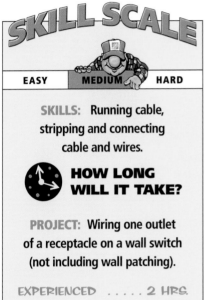

SKILL SCALE

EASY — MEDIUM — HARD

SKILLS: Running cable, stripping and connecting cable and wires.

HOW LONG WILL IT TAKE?

PROJECT: Wiring one outlet of a receptacle on a wall switch (not including wall patching).

EXPERIENCED	2 HRS.
HANDY	3 HRS.
NOVICE	5 HRS.

STUFF YOU'LL NEED

TOOLS: Drill, saw, fish tape, screwdriver, longnose pliers, combination strippers

MATERIALS: Cable and clamps, remodel box, staples, switch, wire nuts, electrician's tape

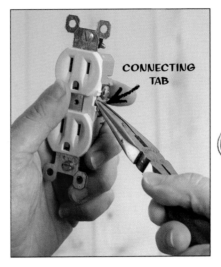

MAKING OUTLETS OPERATE SEPARATELY. Shut off power. In order to make the two outlets of a receptacle operate separately, grasp the connecting tab between the two brass terminals with a pair of longnose pliers. Bend the tab back and forth until it breaks off.

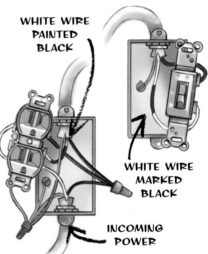

WORKING WITH AN END-OF-THE-RUN RECEPTACLE. Shut off power. Run two-wire cable from the switch to the receptacle. Paint both ends of the white wires black. Connect the grounds. At the receptacle, remove the old black wire and splice it to the new white wire (marked black) and a black pigtail. Connect the pigtail to the always-hot terminal and the other black wire to the other terminal. Attach both wires to the switch.

ALWAYS HOT OUTLET

SWITCHED OUTLET

MIDDLE-OF-THE-RUN RECEPTACLE

SWITCH FOR LOWER OUTLET

END-OF-THE-RUN RECEPTACLE

When you assign one outlet of a duplex receptacle to a wall switch, you can control a floor or table lamp from a doorway. The second outlet will remain hot all the time.

To run cable through finished walls and install a remodel box for the switch, see pages 130–134. See pages 12–13 for grounding methods.

WORKING WITH A RECEPTACLE IN THE MIDDLE OF A RUN. Shut off power. This project will be complicated if the receptacle you want to switch has wires attached to all four terminals. At the receptacle to be switched, remove both old black wires. Splice them with the new white wire and a black pigtail. Connect the pigtail to the always-hot outlet and the new black wire to the switched outlet. Wire the switch and connect the grounds.

SPLITTING A RECEPTACLE

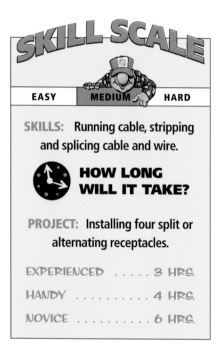

SKILL SCALE

EASY   MEDIUM   HARD

SKILLS: Running cable, stripping and splicing cable and wire.

HOW LONG WILL IT TAKE?

PROJECT: Installing four split or alternating receptacles.

EXPERIENCED . . . . . 3 HRS.

HANDY . . . . 4 HRS.

NOVICE . . . . 6 HRS.

✓ STUFF YOU'LL NEED

TOOLS: Drill, drywall saw or saber saw, fish tape, screwdriver, lineman's pliers, strippers

MATERIALS: Cable and clamps, receptacles, boxes, wire nuts, electrician's tape, cover plates

TIME SAVER

SWITCHING OPTIONS
Install raceway wiring to avoid the hassles of cutting and patching walls (page 138–139). Or, if a fixture's housing is large enough, install an "anywhere" switch (page 98).

Wherever you're likely to plug in more than one high-amp appliance, you run the risk of overloading a circuit. In some cases, two appliances or tools plugged into the same receptacle can add up to more than the circuit can handle.

That's why some building departments require kitchen counter receptacles to be split, so that each outlet is on a separate circuit (right). (In this case, the receptacles cannot be GFCIs.) Other municipalities prefer alternating the receptacles so that every other one is on the same circuit (below). Either of these configurations may be used in a workshop.

Use 15-amp receptacles and breakers and #12 wire for countertops and shop areas.

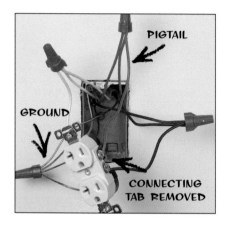

WIRING SPLIT RECEPTACLES. Run three-wire cable, with the black and red wires connected to separate circuit breakers or to the two poles of a double-pole breaker. Break off the connecting tabs on each receptacle (page 143). Using pigtails, connect the white wire to a silver terminal, the red wire to a brass terminal, and the black wire to the other brass terminal. Connect the grounds. Wire the other receptacles the same way.

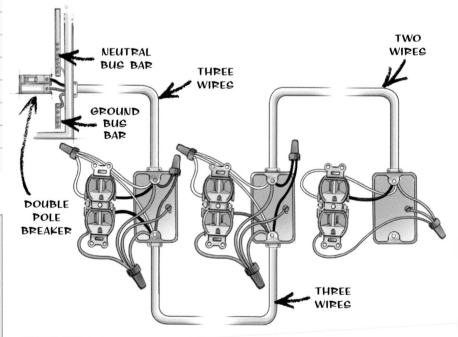

ALTERNATING CIRCUITS. Some codes call for a double-pole breaker, as shown, rather than two separate breakers. When one circuit is turned off, the other is off as well. Use three-wire cable. The black wire brings power to every other receptacle, and the red wire energizes the others. All receptacles share the same neutral wire.

INSTALLING NEW SERVICES

ADDING A 240-VOLT RECEPTACLE

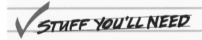

✓ STUFF YOU'LL NEED

TOOLS: Drill, saw, fish tape, screwdriver, lineman's pliers, combination strippers

MATERIALS: 240- or 240/120-volt receptacle, cable, box, wire nuts, electrician's tape

W iring a high-voltage receptacle is slightly more complicated than wiring a standard 120-volt receptacle. Follow safety precautions strictly, however, because this amount of voltage is dangerous.

Choose a receptacle that matches the appliance you will plug into it, both in hole configuration

and amperage rating. Recent codes require four-wire receptacles; three-wire receptacles were once acceptable (page 76-77). Be sure the

wires are thick enough. Use #10 wire for a 30-amp receptacle and #8 wire for a 40- or 50-amp receptacle.

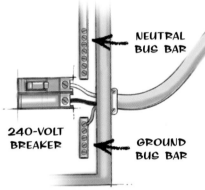

240-VOLT AIR-CONDITIONER RECEPTACLE. A 20-amp, 240-volt receptacle for a window air conditioner or other appliance requires only 12/2 cable, not the heftier #8 wire most 240-volt

receptacles require. Connect the grounds. Mark the white wire black at both ends. Connect the two wires to a 240-volt breaker in the service panel and to the receptacle terminals.

120/240-VOLT RECEPTACLE

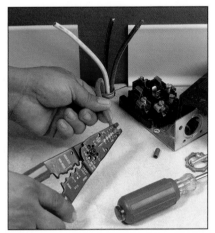

1 RUN THE WIRE. Though some codes allow a 40-amp circuit for a 50-amp range, many electricians prefer a 50-amp circuit so that the range will be protected when all burners are on at the same time (say, at Thanksgiving). Run three-wire cable with #8 wire and strip ¾ inch of insulation from each wire. Some codes require four-wire cable (page 77); check with your building department.

2 INSTALL THE RECEPTACLE. Fasten the receptacle base to the floor or wall with general purpose screws. Connect the black and red wires to the 240-volt breaker, and the white and green to the neutral bus bar. Connect the black, red, and white wires to the receptacle, and connect the ground. Attach the cover housing.

ADDING A FIXTURE WITH SWITCH

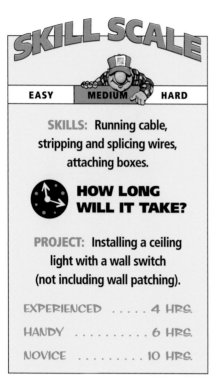

SKILL SCALE

EASY   MEDIUM   HARD

SKILLS: Running cable, stripping and splicing wires, attaching boxes.

HOW LONG WILL IT TAKE?

PROJECT: Installing a ceiling light with a wall switch (not including wall patching).

EXPERIENCED . . . . . 4 HRS.

HANDY . . . . 6 HRS.

NOVICE . . . . 10 HRS.

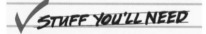

✓ STUFF YOU'LL NEED

TOOLS: Drill, fish tape, screwdriver, lineman's pliers, combination strippers

MATERIALS: Light fixture and box, switch and box, electrician's tape, cable, staples, wire nuts

W hen planning to install a new light with a switch, decide whether to send power into the switch box (above) or to the fixture box (below). Shut off power to the box from which you will run power. See pages 130–134 for instructions on installing boxes and running cable, and pages 12–13 for grounding methods.

POWER TO SWITCH BOX

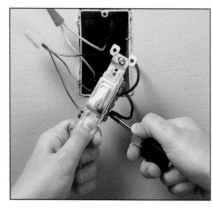

1 **RUN WIRE FROM THE POWER SOURCE TO THE SWITCH BOX.** Run two-wire cable from a power source to a wall switch box, and from there to a ceiling box. Connect the ground. Splice the white wires and connect the black wires to the switch terminals.

2 **WIRE THE FIXTURE.** Connect the ground. Splice the white wire to the white fixture lead, and splice the black wire to the black fixture lead. Secure fixture in place. Restore power and test.

POWER TO FIXTURE

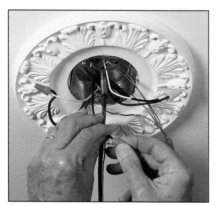

1 **RUN POWER TO THE FIXTURE.** Run two-wire cable from the power source to the fixture box. Run two-wire cable from the fixture box to the switch, and mark the white wire black at both ends. Connect the ground at the fixture box. Splice the black feed wire (from the power source) to the white wire that is marked black. Splice the other black wire to the fixture's black lead, and splice the white wire to the white lead.

2 **WIRE THE SWITCH.** Connect the ground. Connect the black wire and the white wire (painted black) to the switch terminals. Restore power and test.

TWO FIXTURES WITH SEPARATE SWITCHES

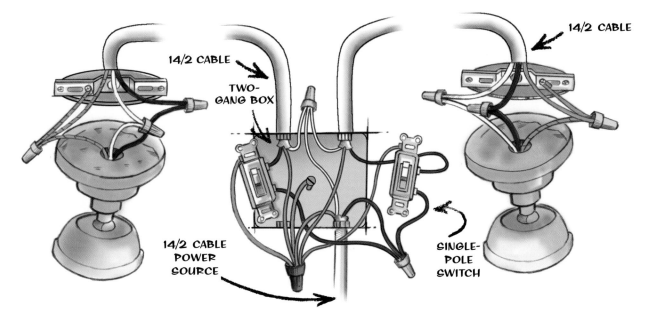

WHEN POWER ENTERS THE SWITCH BOX. Shut off power to the circuit. Run one two-wire cable from the power source into a two-gang switch box, and additional two-wire cables from the switch box to each fixture box. At the switch box, connect the grounds. Splice all the white wires together and splice two black pigtails to the feed wire. Connect one black pigtail and one black wire to each switch. At each fixture box, connect the grounds. Splice the white lead to the white wire and the black lead to the black wire. Restore power and test.

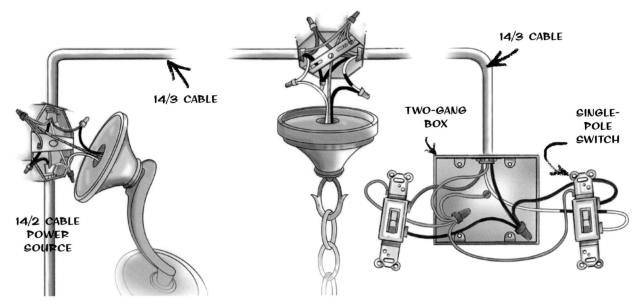

WHEN POWER ENTERS THE FIXTURE. Shut off power to the circuit. Run two-wire cable from a power source to one fixture box. Run three-wire cable from there to the second fixture box. Run three-wire cable from the second fixture box to the two-gang switch box and mark the white wire black at both ends. At the fixture box farthest from the switches, connect the grounds. Splice the two white wires to the fixture's white lead and splice the fixture's black lead to the red wire. Splice the remaining black wires. At the second fixture box, connect the grounds. Splice the marked white wire to the fixture's black lead and the unmarked white wire to the fixture's white lead. Splice the red wires together and splice the black wires together. At the switch box, splice two pigtails to the black wire. Connect the red wire and one pigtail to one switch and the marked white wire and a pigtail to the other switch. Restore power and test.

WIRING THREE-WAY SWITCHES

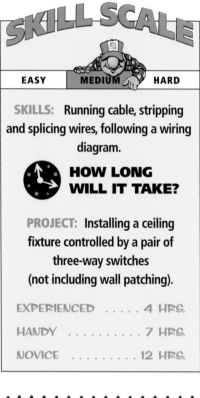

SKILL SCALE

EASY   MEDIUM   HARD

SKILLS: Running cable, stripping and splicing wires, following a wiring diagram.

HOW LONG WILL IT TAKE?

PROJECT: Installing a ceiling fixture controlled by a pair of three-way switches (not including wall patching).

EXPERIENCED ..... 4 HRS.

HANDY .... 7 HRS.

NOVICE .... 12 HRS.

✓ STUFF YOU'LL NEED

TOOLS: Drill, fish tape, lineman's pliers, screwdriver, combination strippers

MATERIALS: Two three-way switches, ceiling fixture, cable and clamps, wire nuts, staples, electrician's tape

INSTALLING NEW SERVICES

hree-way switches are so named because there are three components: two switches and the light fixture. That means you can turn a stairwell light on or off from the top or bottom of the stairs. In long hallways, three-way switches allow you to conveniently control light fixtures from both ends of the hall. In attics, basements, and garages, they spare you from having to grope in the dark looking for the light switch, and they are particularly useful in households with young children who have a tendency to forget to turn off the lights.

See pages 130–134 for tips on running cable and installing boxes and pages 12–13 for grounding methods.

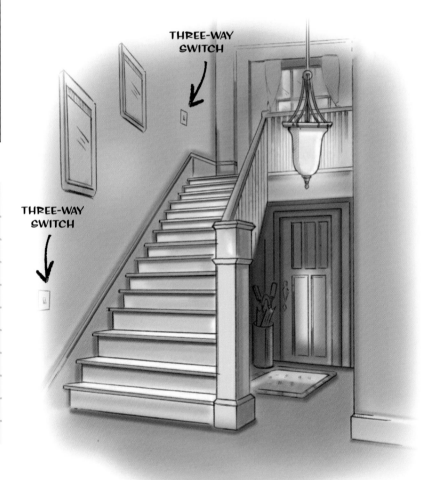

THREE-WAY SWITCH

THREE-WAY SWITCH

THREE-WAY SWITCHES ON STAIRS. Codes—and common sense—dictate that stairway lighting should be controlled by one switch at the bottom of the stairs and one at the top.

CLOSER LOOK

HOW THREE-WAYS WORK

Three-way switches have three terminals. The light they control is turned on when the two switches provide a continuous pathway for power. When either switch creates a gap in that pathway, the light is off.

In a three-way system, a pair of "traveler" wires travel from switch to switch, never to the fixture itself. It is the "common" wire that carries power to the fixture. When you wire a three-way switch, keep in mind that the traveler terminals are interchangeable—it doesn't matter which traveler wire goes to

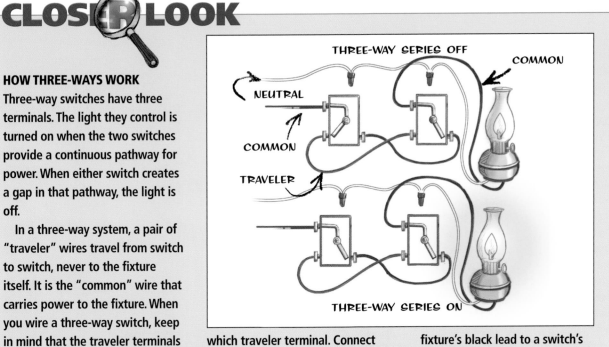

THREE-WAY SERIES OFF

COMMON

NEUTRAL

COMMON

TRAVELER

THREE-WAY SERIES ON

which traveler terminal. Connect either the feed wire (which brings power) or a wire that attaches to the

fixture's black lead to a switch's common terminal.

WHEN POWER RUNS FIXTURE-SWITCH-SWITCH

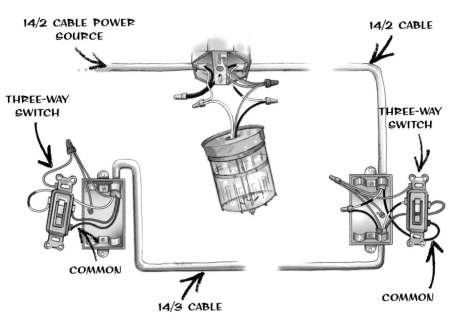

14/2 CABLE POWER SOURCE

14/2 CABLE

THREE-WAY SWITCH

THREE-WAY SWITCH

COMMON

COMMON

14/3 CABLE

WHEN POWER RUNS TO THE FIXTURE FIRST. Shut off power to the circuit. Run two-wire cable from the power source to the fixture box. Run two-wire cable from the fixture box to the first switch box and mark the white wire black

at both ends. Run three-wire cable between the two switch boxes, and mark the white wire black at both ends. Connect the grounds in all three boxes.

At the fixture box, splice the black wires together. Splice the unmarked white wire to the fixture's white lead and splice the marked white wire to the black lead. At the first switch box, connect the marked white wire that comes from the other switch and the red wire to the traveler terminals. Connect the black wire that comes from the fixture to the common terminal. Splice together the remaining wires (one black and one white marked black).

At the second switch box, attach the black-marked white wire and the red wire to the traveler terminals. Attach the black wire to the common terminal. Restore power and test.

WHEN POWER RUNS SWITCH-SWITCH-FIXTURE

WHEN POWER RUNS TO THE SWITCHES FIRST. This is the simplest way to wire three-ways. Shut off power to the circuit. Run two-wire cable from the power source to the first switch box, three-wire cable between the switch boxes, and two-wire cable from the second switch box to the fixture box. Connect the grounds. At the first switch box, splice the white wires. Connect the black feed wire to the common terminal and the other two wires to the traveler terminals. At the second switch box, connect the black wire coming from the fixture to the common terminal (in this case, the lead marked "common" from a three-way dimmer switch). Splice the white wires. Connect the remaining black and red wires to the traveler terminals. At the fixture box, splice the black wire to the black lead and the white wire to the white lead. Restore power and test.

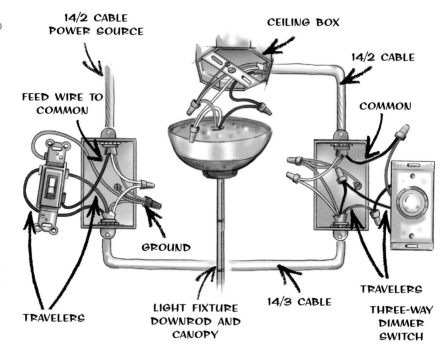

14/2 CABLE POWER SOURCE

CEILING BOX

14/2 CABLE

FEED WIRE TO COMMON

COMMON

GROUND

LIGHT FIXTURE DOWNROD AND CANOPY

14/3 CABLE

TRAVELERS

TRAVELERS

THREE-WAY DIMMER SWITCH

WHEN POWER RUNS SWITCH-FIXTURE-SWITCH

WHEN POWER RUNS SWITCH-FIXTURE-SWITCH. This is the most complicated three-way wiring configuration, but it is sometimes the easiest way to run the cable. Shut off power to the circuit. Run two-wire cable from a power source to the first switch box and three-wire cable from there to the fixture box. Run three-wire cable from the fixture box to the second switch and mark the white wire black at both ends. Connect the grounds in all three boxes. At the first switch box, splice the white wires together. Attach the feed wire to the common terminal and the remaining wires to the traveler terminals. At the fixture box, splice the red wires together. Splice the black wire that comes from the first switch to the white wire that is marked black. Splice the remaining black wire to the black lead and the white wire to the white lead. At the second switch, attach the black wire to the common terminal and the remaining wires to the traveler terminals. Restore power and test.

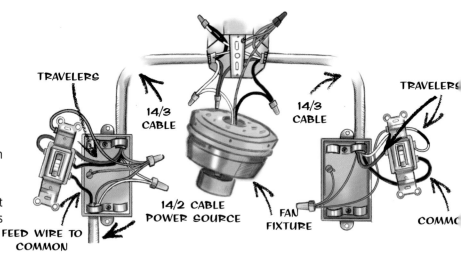

TRAVELERS

14/3 CABLE

14/3 CABLE

TRAVELERS

FEED WIRE TO COMMON

14/2 CABLE POWER SOURCE

FAN FIXTURE

COMMO

WIRING FOUR-WAY SWITCHES

To control a fixture from three or more locations, install a pair of three-way switches at either end and one or more four-way switches in between. The wiring for this setup can get complex, so you may want to hire an electrician. Shown below is a switch-switch-switch fixture; four-way switches also can be wired with power entering the fixture first.

Install as many four-way switches as you like, as long as the first and last switches are three-ways.

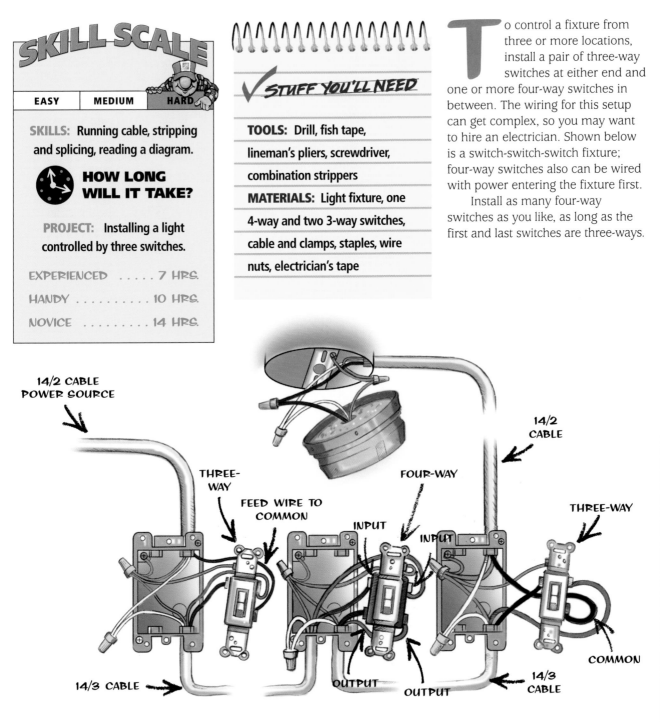

14/2 CABLE POWER SOURCE

14/2 CABLE

THREE-WAY

FEED WIRE TO COMMON

FOUR-WAY

INPUT

INPUT

THREE-WAY

COMMON

14/3 CABLE

OUTPUT

OUTPUT

14/3 CABLE

INSTALLING NEW SERVICES

WIRING A FOUR-WAY SETUP. Shut off power to the circuit. Run two-wire cable from a power source to the first switch box. Run three-wire cable from the first switch box to the second and from the second to the third switch box. Run two-wire cable from the third switch box to the fixture box. Connect all the grounds.

At the first switch box, connect the black feed wire to the common terminal of a three-way switch. Splice the white wires and connect the remaining wires to the traveler terminals. At the second switch box, splice the white wires.

Connect the remaining wires to a four-way switch (which has only traveler terminals, no common terminal), as shown. One set of wires should be on the input terminals and the other set on the output terminals.

At the third switch box, splice the white wires. Connect the black wire that comes from the fixture to the common terminal of a three-way switch and the other two wires to the traveler terminals.

At the fixture box, splice white wire to white lead and black wire to black lead. Restore power and test.

ADDING A WALL LIGHT

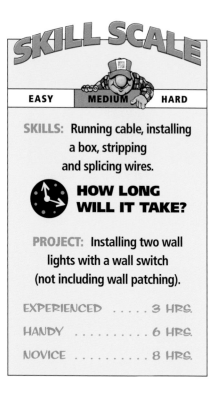

SKILL SCALE

EASY **MEDIUM** HARD

SKILLS: Running cable, installing a box, stripping and splicing wires.

HOW LONG WILL IT TAKE?

PROJECT: Installing two wall lights with a wall switch (not including wall patching).

EXPERIENCED . . . . . 3 HRS.

HANDY . . . . 6 HRS.

NOVICE . . . . 8 HRS.

STUFF YOU'LL NEED

TOOLS: Drill, saw, fish tape, screwdriver, lineman's pliers, combination strippers, level

MATERIALS: Wall sconces or bathroom wall fixture, boxes, cable with clamps, staples, wire nuts, electrician's tape

The methods for installing wall fixtures are the same as those for wiring ceiling lights (page 146). The difference, of course, is that you're working on a vertical surface.

Wall sconces are ideal for hallways and stairwells. Consider wiring them using three-way switches (pages 148–150).

Most wall fixtures attach to a ceiling box. However, check the hardware to make sure you will be able to install the sconce plumb. Buy a swivel strap (page 29) so you can easily adjust the fixture. A fluorescent fixture (for use over a bathroom mirror, for example) may not require a box (page 104).

Homer's Hindsight

GETTING THE RIGHT HEIGHT

To light up my dark hallway, I installed two sconces about 6 feet off the floor. But they made my guests feel like flashlights were shining in their eyes. So I moved the sconces up to 7 feet. Now they provide general illumination—and people can admire them without squinting.

INSTALLING A SCONCE. Run cable from a nearby receptacle or other power source into a switch box and then to a box mounted on the wall. The swivel strap lets you adjust the base until it is level. Depending on the sconce, use either a ceiling fixture box or a switch box. Wire as you would for a ceiling fixture (page 146).

WIRING A VANITY LIGHT. Installing a light over a mirror or medicine cabinet calls for no special wiring techniques. Some fixtures require a box, while others can be wired and then attached directly to the wall. If you will be installing a mirror that reaches to the ceiling, give the glass company exact dimensions for cutting a hole to attach the fixture to a box mounted in the wall behind the mirror.

INSTALLING AN ATTIC FAN

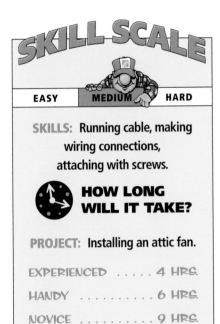

SKILL SCALE

EASY — **MEDIUM** — HARD

SKILLS: Running cable, making wiring connections, attaching with screws.

HOW LONG WILL IT TAKE?

PROJECT: Installing an attic fan.

EXPERIENCED	4 HRS.
HANDY	6 HRS.
NOVICE	9 HRS.

✓ STUFF YOU'LL NEED

TOOLS: Drill, fish tape, screwdriver, lineman's pliers, combination strippers

MATERIALS: Attic fan, cable with clamps, wire nuts, electrician's tape

Temperatures in an attic can reach 150 degrees in the summer, making it difficult (and expensive) to keep a home cool. An attic fan, a whole-house fan (page 154–155), or a roof fan (page 155) slashes energy costs and reduces temperatures.

The manufacturer should provide a chart detailing how powerful a fan you need based on the size of your attic. Depending on the size of your house, you may require more than one fan.

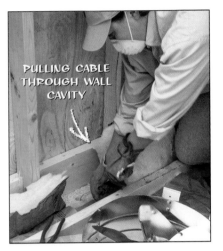
PULLING CABLE THROUGH WALL CAVITY

1 BRING POWER INTO THE ATTIC. Before tapping into a receptacle or junction box for power, check the amperage on your attic fan and make sure you will not overload the circuit (pages 116–117). Shut off power to the circuit. See pages 130-132 for tips on running cable into the attic. Check with local codes to see whether you need to use armored cable instead of NM cable.

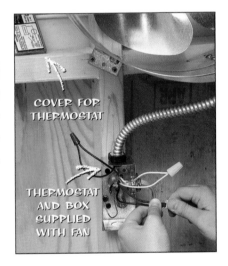
COVER FOR THERMOSTAT

THERMOSTAT AND BOX SUPPLIED WITH FAN

3 MAKE THE ELECTRICAL CONNECTIONS. The fan has its own thermostat switch. Mount the thermostat box to a framing member. Follow the manufacturer's instructions for connecting wires. Restore power, and adjust the temperature control. Or, you can control the fan with a pilot-light switch (page 98) located in the hallway.

2 MOUNT THE FAN. At a louvered opening in the attic, secure the fan by driving screws through its mounting brackets and into studs. If the studs do not allow you to center the fan in the opening, attach horizontal 2×4s that span between the studs. Attach the fan to the studs. You may choose to install louvers that close when the fan is not operating.

CLOSER LOOK

AN ATTIC MUST BREATHE

An attic fan, whole-house fan, or roof fan moves air efficiently only if the attic is properly ventilated. Usually, a house needs vents near the bottom of the attic (usually under the eaves) and vents near the roof peak, such as turbine vents, gable vents, or a continuous vent running along the ridge. Check that eave vents are not clogged with insulation. If you are not sure that your attic is properly vented, have it inspected by a professional roofer.

153

INSTALLING A WHOLE-HOUSE FAN

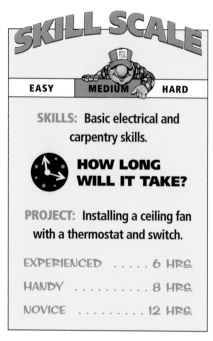
STUFF YOU'LL NEED

TOOLS: Drill, ladder, saw, combination strippers, lineman's pliers, screwdriver

MATERIALS: Whole-house fan, screws, junction box, switch box, cable with clamps, wire nuts, electrician's tape

Find a powerful yet quiet whole-house fan to pull up air through the house and into your attic. A whole-house fan is ideal for spring and fall cooling in hot climates; it may be the only means of cooling you need in moderate climates. For the fan to work, the attic must have adequate ventilation (page 153) and windows on the first floor must be open. Measure your home's square footage to choose the right size fan.

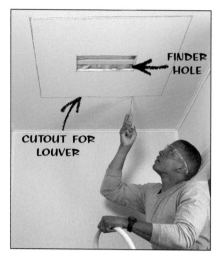

FINDER HOLE

CUTOUT FOR LOUVER

1 CUT A HOLE. Fans are designed to be positioned over one joist so you don't have to compromise ceiling framing. Cut a 1×2-foot finder hole to confirm that the fan will center on a joist. Mark the cutout for the louver and cut through the drywall or lath and plaster. The fan manufacturer will specify dimensions for the hole.

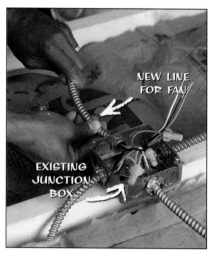

NEW LINE FOR FAN

EXISTING JUNCTION BOX

2 BRING POWER TO THE FAN. After making sure that you will not overload a circuit (pages 116–117), shut off power to the circuit. Tap into a junction box or run cable up into the attic (see pages 130–132). Local codes may require you to use armored cable instead of NM cable.

NM CABLE TO HALLWAY SWITCH

3 WIRE FOR THE SWITCH. A fan-rated rheostat switch lets you vary the fan speed. Bring the two-wire switch cable to the box, marking the white wire black at both ends. Splice it to the black wires in the box. Splice the other switch wire to the fan's black wire and the fan's white to the white wires in the box. Connect the ground.

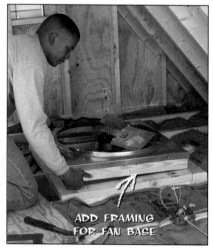

ADD FRAMING FOR FAN BASE

4 MOUNT THE FAN. With a helper, lift the fan up through the opening and into the attic. Add framing as needed so the fan is securely centered over a joist. Attach brackets to the fan frame and position them so they will slip over the exposed joist. Center the fan over the opening and secure the brackets with bolts.

INSTALLING NEW SERVICES

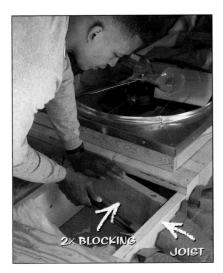

5 **ENCLOSE THE FAN.** Pull back the insulation and cut pieces of 2× blocking to fill gaps at either side of the fan. At each side, cut two pieces to fit between the joists (shown) or one notched piece that fits over the joist. (Some manufacturers supply blocking.) Attach the wood to the joists by drilling pilot holes and attaching with 3-inch screws or 16-penny nails.

6 **WIRE THE RHEOSTAT SWITCH.** Install a fan-rated rheostat switch in the hallway, connecting it to the cable you have run through the wall from the attic junction box. Wire as shown or use the manufacturer's directions.

7 **ATTACH THE LOUVERS.** Hold the louver panel against the ceiling so it covers the hole. Attach the panel by driving screws into the joists and blocking. Restore power and test the fan.

CLOSER LOOK

A FAN TO PULL THE AIR OUT OF YOUR ATTIC

If your attic does not have a vertical wall to accommodate an attic fan (page 153), this is the next most efficient way to pull air out of an attic. The most difficult part of this installation is not the electrical work, but the roofing. The shingles must be laid correctly over the fan flashing, or the roof will leak. If you have an existing passive roof vent the same size as your fan, you can install the fan in that spot without much trouble. Call a roofer if you aren't sure how to seal the fan.

1 **CUT THE HOLE AND ROOFING.** Follow the manufacturer's directions for cutting a hole through the roof and for cutting back shingles from around the hole. Carefully fold back the shingles.

2 **INSTALL THE FAN.** Slide the fan under the shingles and apply roofing cement as directed. Wire the fan as you would an attic fan (page 153).

INSTALLING NEW SERVICES

155

INSTALLING A BATHROOM VENT FAN

A vent fan considerably improves the atmosphere in a bathroom by pulling out moisture, odors, and heat. Codes require bathrooms to have vent fans if there is no natural ventilation, such as a window. You may opt for a fan even if you have a window, so you can clear the air in rainy or cold weather. When choosing a fan, use these guidelines:

- **Make sure the fan will move the air.** Unfortunately, many bathroom fans do little more than make noise. This happens when either the fan is not strong enough or the path through the ductwork is not free and clear. Measure your room and determine how far the ductwork has to travel. Then ask a home center salesperson to help you choose a fan and the ductwork to do the job (page 158). Keep in mind that air travels more freely through solid ducts than through flexible hoses.
- **Consider a fan equipped with a light.** Some units have a fan only, while others include a ceiling light, a low-wattage night-light, and even a forced-air heating unit.
- **Consider the wiring options.** At times, local codes require that the fan come on whenever the overhead light is turned on. Some people prefer separate switches for the bathroom fan and light.

INSTALLING DUCTWORK

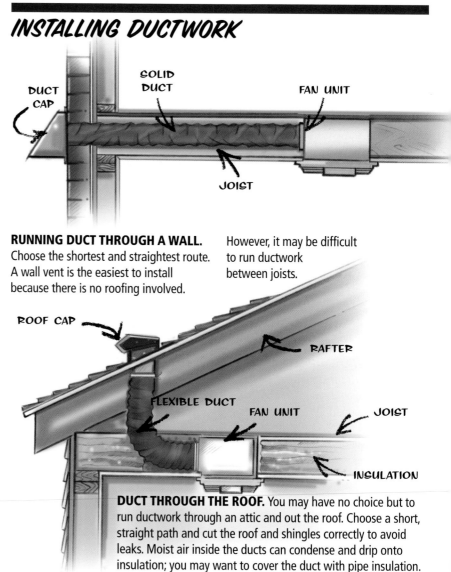

RUNNING DUCT THROUGH A WALL. Choose the shortest and straightest route. A wall vent is the easiest to install because there is no roofing involved. However, it may be difficult to run ductwork between joists.

DUCT THROUGH THE ROOF. You may have no choice but to run ductwork through an attic and out the roof. Choose a short, straight path and cut the roof and shingles correctly to avoid leaks. Moist air inside the ducts can condense and drip onto insulation; you may want to cover the duct with pipe insulation.

1 **CUT THE HOLE.** From the attic above, hold the fan against a joist and mark its outline with a pencil. Cut out the opening. If there is no attic above, use a stud sensor to locate a joist and cut the opening from below. Shut off power to the circuit and provide power if none is present (pages 128–132).

2 **ATTACH THE FAN AND DAM OFF INSULATION.** Attach the fan to the joist with screws. Some models require a 6-inch gap between the unit and insulation. Cut or push back the insulation; then cut pieces of 2× lumber to fit between the joists and attach the lumber with screws or nails.

3 **CUT A HOLE IN THE ROOF.** On the underside of the roof, trace a circle just large enough for the roof cap tailpiece. Drill a hole large enough for the saw blade, then cut with a reciprocating saw, saber saw, or keyhole saw. (If you run the ductwork out the wall, see page 158.)

4 **CUT AWAY SHINGLES.** Remove shingles from around the cutout without damaging the underlying roofing paper. The lower part of the roof cap flange will rest on top of the shingles, and the top part will slip under the shingles.

5 **INSTALL THE ROOF CAP.** Smear roofing cement on the underside of the cap flange. Slip the upper flange under the shingles as you insert the cap into the hole. Install the shingles on the side, smearing the undersides with roofing cement. Attach the flange with roofing nails and cover the heads with roofing cement.

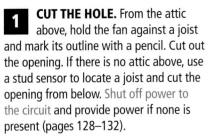

BUYER'S GUIDE

NOISE CONTROL

A label on the fan packaging will indicate how many square feet of bathroom space the fan can successfully clear. If there's any doubt, or if your ductwork will be more than 5 feet long, get a slightly more powerful fan than you need. (However, don't overdo it. Keep the power of the fan appropriate to the size of the room.) The SONE rating on a fan indicates its sound-level rating. A 3 SONE rating is quiet; a 7 will be very noisy.

INSTALLING NEW SERVICES

FLEXIBLE DUCT

6 CONNECT THE DUCTWORK.
Flexible ductwork is the easiest to run, but solid ducts are quieter and more efficient. At both the roof cap and the fan, slide a clamp over the duct and slip the duct over the tailpiece. Slide the clamp back over the tailpiece and tighten the clamp. Wrap the joint with duct tape.

7 WIRE THE FAN. If wiring does not exist, run cable to the fan and to a switch. If you are installing a fan/light, run three-wire cable from the switch to the fan. Connect the wiring according to the manufacturer's directions. Plug the motor into the built-in receptacle.

8 WIRE THE SWITCH. For a fan/light switch that has power entering the switch box, splice the white wires and connect the grounds. Connect power to both switches through two pigtails spliced to the feed wire. Connect the red wire to one switch terminal and the black wire to the other terminal.

A+ WORK SMARTER

INSTALLING A WALL VENT

If a bathroom is not located directly beneath an attic, you must vent air out through a wall. Even if there is an attic above, it may be easier to run the vent out through a gable wall rather than through the roof. When running ductwork through a ceiling cavity, it is sometimes easier to shove a piece of solid ductwork through rather than snaking flexible ducting. From inside the attic, drill a locator hole through to the outside, then cut out the siding with a reciprocating saw, saber saw, or keyhole saw.

DUCT PIPE

DUCT CAP

1 MAKE A TAILPIECE. Press the duct pipe into the cap. Use sheet-metal screws to attach a piece of solid duct to the cap, then caulk the joint or wrap it with duct tape. Apply a bead of caulk to the back of the flange so it will seal against the siding.

2 ATTACH THE VENT. Caulk around the hole and push in the tailpiece. Secure it with four screws. Caulk around the edge of the vent. Complete the connection to the fan indoors using solid or flexible ductwork.

ADDING A RANGE HOOD

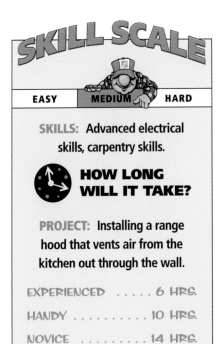

SKILL SCALE

EASY | **MEDIUM** | HARD

SKILLS: Advanced electrical skills, carpentry skills.

HOW LONG WILL IT TAKE?

PROJECT: Installing a range hood that vents air from the kitchen out through the wall.

EXPERIENCED . . . . . 6 HRS.

HANDY . . . . 10 HRS.

NOVICE . . . . 14 HRS.

✓ STUFF YOU'LL NEED

TOOLS: Drill, fish tape, saber saw or reciprocating saw, hammer and cold chisel, screwdriver, combination strippers, lineman's pliers

MATERIALS: Range hood, solid duct, wall cap, masonry screws, cable and clamps, wire nuts, electrician's tape, caulk, safety goggles

Most residential range hoods, if correctly installed, remove smoke, odor, and heat from the kitchen. To draw out cooking grease, you need a powerful commercial model.

For the best range hood efficiency, run the duct through the wall directly behind the range hood, in as straight a line as possible. You can run the vents of most hoods out the back or the top of the unit.

If a wall stud is in the way of the ductwork, you could do carpentry work to change the framing. An easier solution is to purchase a hood with an extra-strong motor and run the duct around the stud.

Before you purchase a fan, check its "cfm" rating—which indicates the number of cubic feet of air it pulls per minute. Choose a fan with a cfm rating that is double the square footage of your kitchen.

TIME SAVER

DUCTLESS RANGE HOODS

Cutting a hole in the wall and running ductwork is the most time-consuming and difficult part of installing a range hood. Save yourself the hassle by installing a ductless hood, which runs air through a filter and back into the kitchen rather than moving it outside. This unit will not be as effective at removing smoke and odors, however, and you'll need to change filters fairly often.

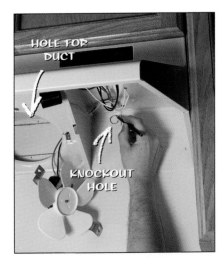

1 **MARK FOR HOLES.** Remove the filter, fan, and electrical housing cover from the range hood. Use a hammer and screwdriver to remove the knockouts for the electrical cable and the duct. Hold the hood in place and mark the holes for the duct and the cable.

2 **CUT THE INSIDE AND DRILL A LOCATOR HOLE.** Cut holes through the drywall or plaster. Using a long bit, drill holes at each corner all the way through the outside wall. (If your exterior is brick or block, see page 160.)

INSTALLING NEW SERVICES

3 **CUT THE SIDING.** Connect the dots between the holes on the outside to mark the outline of the hole. Using a reciprocating saw, saber saw with an extra-long blade, or keyhole saw, cut the outline. Remove insulation or debris that would interfere with installing the duct.

4 **ATTACH THE DUCT CAP.** Push the wall cap into the wall to see if the duct is long enough to reach the range hood. If not, purchase an extension and attach it with sheet-metal screws and duct tape. Apply caulk to the siding where the cap flange will rest. Push the cap into place and fasten with screws. Caulk the perimeter of the flange.

5 **RUN POWER TO THE HOOD.** Shut off power to the circuit. Run cable from a nearby receptacle or junction box through the hole in the wall (pages 130–132). Strip the sheathing and clamp the cable to the range hood electrical knockout. Mount the hood securely by driving screws into studs or adjacent cabinets.

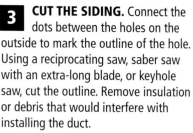

6 **CONNECT THE WIRES.** Splice the white wire to the white fixture lead, black wire to black lead, and the ground wire to the green lead. Fold the wires into place and replace the electrical cover. Reattach the fan and filter. Restore power and test.

CLOSER LOOK

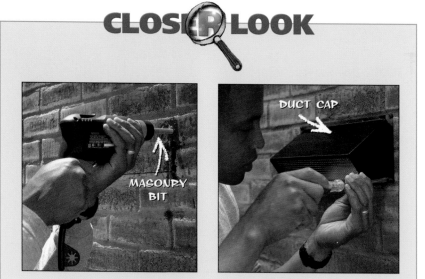

VENTING THROUGH A MASONRY WALL

Use a long masonry bit to drill the locator holes (see Step 2, page 159). Draw the outline carefully, double-checking that you can slip in the vent with room to spare. Drill holes about every inch along the outline; then use a hammer and cold chisel to chip between the holes. To attach the duct cap, drill holes and drive masonry screws. Older homes may have double-thick brick walls—a real challenge!

INSTALLING A SMOKE DETECTOR

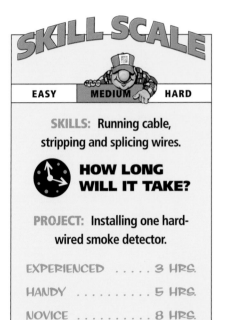

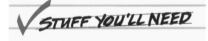

✓ STUFF YOU'LL NEED

TOOLS: Drill, fish tape, combination tool, lineman's pliers, screwdriver

MATERIALS: Smoke detector (hard-wired with battery backup), cable with clamps, ceiling box, wire nuts, electrician's tape

Many homes have battery-powered smoke detectors that fail to perform when the batteries die. Other homes have hard-wired detectors that don't work if the wiring gets damaged—which often happens in a fire. For the best protection, install hard-wired detectors that have battery backup.

Most codes allow you to install battery-only detectors. These are fine as long as you test them regularly and immediately replace failing batteries.

Many detectors are wired with two-wire cable. As an added safety precaution, install detectors in a series, using three-wire cable. That way, when one is triggered, all will sound. (See pages 128–132 for running cable and pages 133–134 for installing remodeling boxes.)

WHERE TO PUT DETECTORS
If your detectors were installed more than 5 years ago, chances are you have too few to satisfy current codes. For instance, detectors are now required both in the hall and inside bedrooms to warn you in case of a bedroom fire. Ask your building or fire department for recommendations.

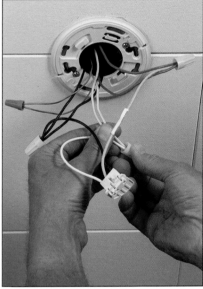

WIRING FOR A SERIES OF DETECTORS. Run three-wire cable to all the detectors so that when one senses smoke, they all screech in unison. (Detectors wired with two-wire cable work independently of one another.) Wire each as shown, following manufacturer's directions.

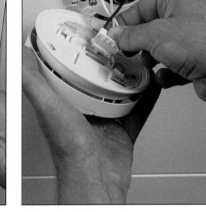

INSTALLING A DETECTOR. Carefully pack the spliced cables into the ceiling box. Clip the connector onto the back of the detector and install the unit.

INSTALLING NEW SERVICES

161

ADDING AN OUTDOOR RECEPTACLE

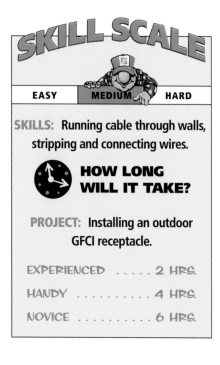

SKILL SCALE

EASY   MEDIUM   HARD

SKILLS: Running cable through walls, stripping and connecting wires.

HOW LONG WILL IT TAKE?

PROJECT: Installing an outdoor GFCI receptacle.

EXPERIENCED . . . . . 2 HRS.

HANDY . . . . 4 HRS.

NOVICE . . . . 6 HRS.

✔ STUFF YOU'LL NEED

TOOLS: Drill with long bit, sabersaw or keyhole saw, combination strippers, screwdriver, lineman's pliers, hammer, (for masonry walls, a masonry bit and cold chisel)

MATERIALS: GFCI receptacle, cable with clamps, remodel box, wire nuts, electrician's tape

Unless you need an outdoor receptacle in a particular location, plan the easiest path for the cable. One option is to install it nearly (but not exactly) back-to-back with an indoor receptacle. Or, run cable through the basement ceiling and out the rim joist.

Even if you install a weatherproof cover, place the receptacle in a dry location and at least 16 inches above the ground. Codes require that an outdoor receptacle be a ground-fault circuit interrupter (GFCI).

Ensure that you will not be overloading a circuit when installing the outdoor receptacle (pages 116–117). If you plug in too many Christmas lights or plan to use heavy-duty power tools, you may need to place the receptacle on its own circuit (pages 168–169). Local codes may require that you have a separate circuit for outdoor electrical service. See pages 130–132 for tips on fishing cable through walls.

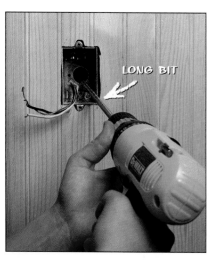

LONG BIT

1 DRILL A LOCATOR HOLE. Shut off power to the circuit. Pull out an interior receptacle and detach it. Using a hammer and screwdriver, open a knockout hole in the back of the receptacle box. Put a long bit or a bit extension in your drill. (Use a masonry bit if your exterior is brick.) Poke the bit through the hole in the box. The wall may not be thick enough to fit back-to-back receptacles, so aim the drill bit at an angle. Drill through to the outside.

KEYHOLE SAW

2 CUT THE HOLE FOR THE RECEPTACLE. On the outside, cut a hole for a receptacle box. Drill a second hole as an entry point and use a saber saw with an extra-long blade, a reciprocating saw, or a keyhole saw. If the exterior is masonry, see page 160 for tips on cutting the hole.

REMODEL BOX

3 **RUN THE CABLE.** Cut cable about 2 feet longer than you need, and strip the sheathing from both ends. Have a helper push the cable from indoors as you pull it out through the hole. Clamp the cable to the remodel box and mount the box (pages 133–134). Install a new GFCI receptacle outside (page 27). Connect to power in the interior box.

IN-USE COVER

4 **INSTALL AN IN-USE COVER.** This kind of cover will keep the receptacle dry even when it has a cord plugged into it, and it can be locked shut. In addition to the plastic cover, install the rubber gasket behind the plate. Restore the power and test the receptacle.

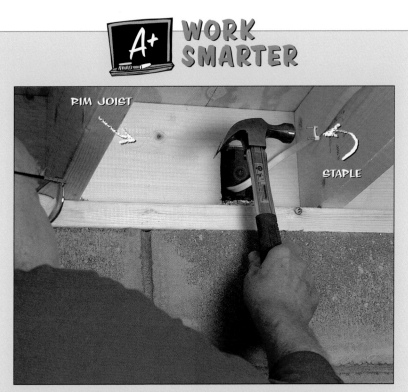
RIM JOIST   **STAPLE**

RUNNING CABLE THROUGH A BASEMENT WALL

If your basement ceiling is unfinished, this is probably the easiest method. Shut off power to the circuit and tap into a receptacle or junction box. Cut a hole through the rim joist and siding, and staple the cable to the joist.

EXTENDING SERVICE FROM AN OUTDOOR RECEPTACLE

Adding an extension ring to the receptacle box allows you to run cable or conduit from the receptacle to supply outdoor lights and other receptacles. Two screws attach the extension ring to the box. Remove one or more knockouts in the ring to extend the circuit.

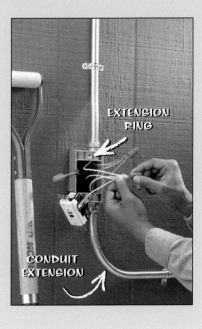
EXTENSION RING

CONDUIT EXTENSION

INSTALLING NEW SERVICES

RUNNING AN OUTDOOR LINE

SKILL SCALE

EASY	MEDIUM	HARD

SKILLS: Advanced wiring, excavating and carpentry.

HOW LONG WILL IT TAKE?

PROJECT: Installing 50 feet of outdoor line, three lights.

EXPERIENCED . . . . . 10 HRS.

HANDY . . . . 12 HRS.

NOVICE . . . . 14 HRS.

✓ STUFF YOU'LL NEED

TOOLS: Spade, post-hole digger, hammer, screwdriver, lineman's pliers, combination tool, hacksaw, crescent wrench, drill

MATERIALS: Rigid metal conduit or PVC and fittings, wire or UF cable, string, stakes, outdoor boxes, 4×4 post, concrete

BUYER'S GUIDE

CONDUIT OPTIONS

Outdoor conduit must be watertight. Use rigid metal conduit with threaded fittings or schedule 40 PVC plastic conduit. (Glue the PVC with cement approved by your local building department.) Take a sketch of your wiring to your supplier for help gathering the parts.

INSTALLING NEW SERVICES

Low-voltage lights (page 40) are quick and easy to install, but if you need outdoor receptacles or strong lighting, roll up your sleeves and prepare to dig trenches and run cable or conduit.

OUTDOOR WIRING OPTIONS

Check with your building department to learn the requirements for outdoor wiring. In some areas, UF (underground feed) cable is fine; elsewhere you must run wires through conduit. Codes also specify whether to use metal or PVC conduit. Codes specify how deep the wiring must be buried and whether you must protect the cable with a 2×6 plank. Conduit usually has to be 18 inches deep and cable must be 24 inches deep.

If you want to install continuous conduit (instead of the conduit-and-cable system shown on the next page), run individual wires—rather than cable—through the conduit.

If you will install only a couple of lights and a receptacle, you may be able to tap into an existing circuit (pages 116–117). However, it is often nearly as easy to run a new circuit for outdoor service (pages 168–169). Your building department may require a separate circuit. If you run more than 50 feet of cable, use #12.

Before digging, call the toll-free numbers for utilities to make sure you won't hit a power, water, gas, phone, or cable line.

1 DIG THE TRENCH. Use a string line to mark the path of the underground cable. Cut the sod neatly and place it right-side-up on the ground so you can reuse it later. Dig to the depth required by codes. If you have a lot of digging to do, consider renting a power trencher. Codes may allow the trench to be more shallow if you cover the cable with a 2×6 plank.

2 START WITH AN LB FITTING. You can begin service with a receptacle (pages 162–163) or with an LB fitting, which makes pulling the wire easier. Start with a short length of conduit sticking out of the house. Attach the LB fitting to it, then a length of conduit downward, then a sweep (a curved piece of conduit) into the trench. Do not splice wires in this fitting.

3 **TAP CONDUIT UNDER SIDEWALK.** Use this method whether running cable or conduit. Cut a piece of rigid metal or PVC conduit 2 feet longer than the sidewalk width and flatten one end with a hammer to make a sharp point. Drive the pipe under the sidewalk with a hammer and a block of wood. If required, install conduit for the entire run. In many areas, however, you can run underground-feed (UF) cable.

4 **CUT OFF THE SHARP END OF THE CONDUIT WITH A HACKSAW.** When using metal conduit, install a plastic bushing at the end of the sweep so the cable won't get nicked as it's pulled through the conduit.

5 **ANCHOR A POST FOR A FREESTANDING LIGHT.** Use a clamshell digger to dig a post hole about 3 feet deeper than the trench. Insert a pressure-treated 4×4 post (make sure it is about 1 foot longer than you need), check it for plumb in both directions, and brace it with 1×4s from two sides. Fill the hole with concrete or well-tamped soil. Once the concrete has set, trim the post to the desired height. Continue your run of underground cable or conduit. Either way, finish with metal conduit, running a sweep out of the trench and up the post to the box.

6 **INSTALL A BOX.** Attach a weatherproof box to the conduit near the top of the post. Mount the conduit to the post with clamps and screws. Run cable into the box or fish wires through the conduit (page 127). Install a GFCI receptacle (page 27).

INSTALLING OUTDOOR LIGHTING

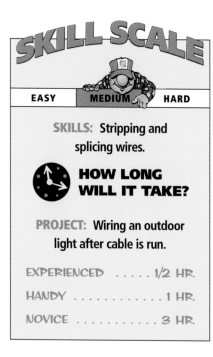

SKILL SCALE

EASY   MEDIUM   HARD

SKILLS: Stripping and splicing wires.

HOW LONG WILL IT TAKE?

PROJECT: Wiring an outdoor light after cable is run.

EXPERIENCED . . . . . 1/2 HR.

HANDY . . . . 1 HR.

NOVICE . . . . 3 HR.

✓ STUFF YOU'LL NEED

TOOLS: Screwdriver, lineman's pliers, combination tool, drill, saber saw, router

MATERIALS: Outdoor light fixture, photocell switch, box

Final connections for outdoor fixtures are about the same as for indoor fixtures. The main difference is that the outdoor components are heftier and have watertight gaskets.

Some lights come with their own posts, and for others you have to get 4×4 posts separately. Others attach to the tops of railings, under eaves, or to the sides of posts. Incandescents and halogens are typical for residential use; mercury-vapor and metal halide lights are more suited for commercial use. High-output fluorescent tubes operate down to minus 40C; screw-in fluorescent bulbs will work down to minus 23C.

Make sure each light's compression fitting fits over the metal or plastic conduit. You may need to buy an adapter to ensure a tight fit. Shut off power to the circuit before connecting wires.

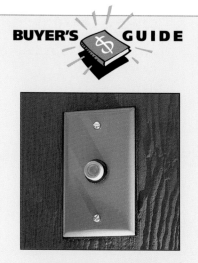

BUYER'S GUIDE

PHOTOCELL SWITCH

This switch turns on a light at dusk and off at dawn. Position the photocell where it can sense daylight and out of the path of artificial light—including the lights controlled by the photocell.

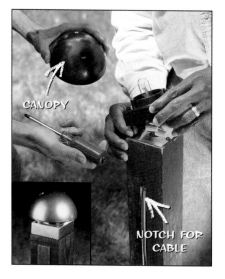

CANOPY

NOTCH FOR CABLE

ADDING A POST LIGHT. The subtle illumination (inset) of this 120-volt light illuminates paths and decorates decks. Notch the post with a router and bore a hole from the top to meet the notch. Run cable to the base, wire it white to white and black to black. Pigtail the ground to the base. Install the canopy and cover the notch with a piece of lattice.

WIRING AN UNDER-EAVE LIGHT. Cable for this kind of light sometimes can be run through the attic. Cut a round hole in the eave, run cable into a remodel box, and attach the box. If you want to install a motion-sensor light (page 108), ask a salesperson to recommend one that will work well on a horizontal surface.

The installations described in this chapter involve adding new circuits. You may want to hire a pro to complete most of these projects, or may have to because your local building department may not allow unlicensed homeowners to run new circuits, install panels or subpanels, or wire entire rooms. If you hire out the work, use this chapter to understand what is involved and how to judge the quality of the work being done.

If you attempt the projects yourself, do so only after you've learned how electricity works and how to calculate loads, and have successfully completed several electrical installations.

Before attempting any of the projects in this chapter, be sure you have a thorough understanding of the wiring principles presented in Chapter 1 and have successfully completed several wiring projects.

CHAPTER ELEVEN PROJECTS

MAJOR PROJECTS

ADDING A NEW CIRCUIT

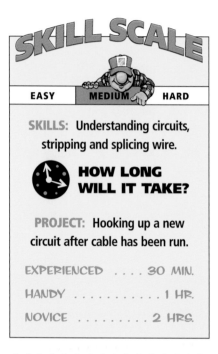

SKILL SCALE

EASY   MEDIUM   HARD

SKILLS: Understanding circuits, stripping and splicing wire.

HOW LONG WILL IT TAKE?

PROJECT: Hooking up a new circuit after cable has been run.

EXPERIENCED	30 MIN.
HANDY	1 HR.
NOVICE	2 HRS.

STUFF YOU'LL NEED

TOOLS: Hammer, screwdriver, lineman's pliers, combination strippers

MATERIALS: Cable and clamp, new circuit breaker

FIRE PROTECTION

Beginning January 2002, the CEC will require the use of Arc Fault Circuit Interrupters (AFCI) for bedroom circuits. AFCIs provide greater fire protection than a regular breaker. Regular breakers trip for overloads and short circuits. AFCIs offer protection when arcing occurs because of frayed and overheated cords, and impaired wire insulation.

The physical work of installing a new electrical circuit is simple and calls for no special skills. Most of the work is completed outside the service panel. To get a breaker that will fit in your panel, jot down the brand and model number, or bring a sample breaker to the store.

First, determine whether your service panel can accommodate a new breaker, and then plan a circuit that will not be overloaded (pages 116–117). Install the new boxes. Run cable from the boxes back to the service panel (pages 128–132). (Electricians call this practice a "home run.") Hook up the devices and fixtures. Now you're ready to energize the new circuit by installing a new breaker and connecting the wires to it.

1 **SHUT OFF MAIN POWER.** Work during the daytime and have a reliable flashlight on hand. Turn off the main circuit breaker. All the wires and circuit breakers in the panel are now de-energized except for the thick wires that come from the outside and connect to the main breaker. **Do not touch them.**

CLOSER LOOK

BREAKER OPTIONS. If the service panel has room, install full-size, **single-pole** breakers. If you're out of space, see if your panel can accommodate **tandem** breakers. In some panels, these breakers only fit in slots near the bottom. Your building department limits the number of breakers that can be installed. If you add too many, they will require you to put in a new panel or a subpanel (pages 170–171). You'll need double-pole breakers for 240-volt circuits. A **quad** breaker can supply two 240-volt circuits.

MAJOR PROJECTS

168

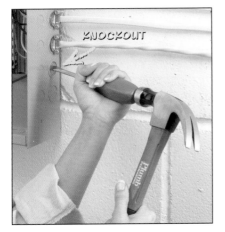

2 **REMOVE A KNOCKOUT.**
Remove the service panel cover (page 14). Remove a knockout slug from the side of the service panel and install a cable clamp. Also, remove a knockout tab from the panel cover. (For a double-pole breaker, remove two knockouts.)

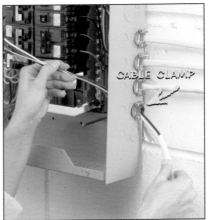

3 **CLAMP THE CABLE.** Determine how far the wires must travel to reach the breaker and the neutral bus bar. To avoid tangles, plan a path around the box perimeter. Strip about a foot more sheathing than you think you need. Thread the wires through the clamp and secure the cable. Don't overtighten.

4 **CONNECT THE NEUTRAL WIRE.**
Run the neutral wire toward an open terminal in the neutral bus bar, bending the wire carefully so it will easily fit behind the panel cover. Cut the wire to length and strip off about ½ inch of insulation. Poke the end into the terminal and tighten the setscrew. Connect the ground wire to the ground bar (or neutral bar if there is no ground bar).

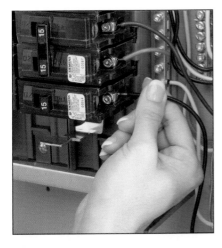

5 **WIRE THE NEW BREAKER.** Run the hot wire, bending it carefully so it will easily fit behind the panel cover. Cut the wire to length. Strip off ½ inch of insulation. Poke the wire into the new breaker terminal. If bare wire is visible, remove the wire, snip it a little shorter, and reinsert it. Tighten the setscrew.

6 **SNAP THE BREAKER INTO PLACE.** Slip one side of the breaker under a tab to the right or left of the hot bus bar. Push the other side onto the bus bar until the new breaker is flush with the other breakers. (Some brands of breakers may require a slightly different installation method. Check the instructions.) Restore power and test.

INSTALLING DOUBLE-POLE BREAKERS. Shut off the power. Wire a 240/120-volt circuit with the black and red wires connected to the breaker terminals. Connect the white wire to the neutral bus bar. Wire split circuits the same way (page 144). To wire a straight 240-volt circuit (page 145), connect the two hot wires to the breaker and the ground wire to the ground bar (or neutral bar if there is no ground bar).

INSTALLING A SUBPANEL

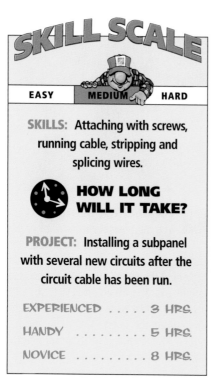

SKILL SCALE

EASY   MEDIUM   HARD

SKILLS: Attaching with screws, running cable, stripping and splicing wires.

HOW LONG WILL IT TAKE?

PROJECT: Installing a subpanel with several new circuits after the circuit cable has been run.

EXPERIENCED . . . . . 3 HRS.

HANDY . . . . 5 HRS.

NOVICE . . . . 8 HRS.

✓ STUFF YOU'LL NEED

TOOLS: Drill, hammer, screwdriver, combination strippers, lineman's pliers

MATERIALS: Subpanel, screws, cable with clamps, staples

An experienced homeowner can tackle a subpanel, but hire a pro if a new service panel is needed (page 172).

WHEN TO ADD A SUBPANEL

Install a subpanel to handle new circuits if the existing service panel does not have open breaker slots and you cannot use half-size breakers (page 168).

A subpanel doesn't add to the total amount of power entering the home. If you have 100-amp service and the new circuits you are installing will require more than that (see page 62 for how to add up your requirements), call an electrician. Electricians working with the utility company can bring 200-amp service to your service head and can install a new service panel.

Purchase a subpanel. Like a main service panel, it has separate bus bars for neutral and ground wires.

Figuring the size of the subpanel, feeder cable, and feeder breaker can be complicated, so consult an electrician or your building department.

In most cases, to add up to six new circuits with a total of 6000 watts or less (pages 116–117), you'll need a 30-amp, 240-volt subpanel. Open two spaces in the main panel and install a 30-amp double-pole feeder breaker. Run 10/3 feeder cable between the main panel and the subpanel. Or install a 40-amp subpanel and feeder breaker, and use #8 wire. Once your plan is approved, get a permit.

1 MOUNT THE SUBPANEL.
Position the subpanel for easy access but out of reach of small children. As with a main panel, there must be drywall or other nonflammable material between the subpanel and wood framing. Anchor it firmly by driving screws into studs. On a masonry wall, drill holes and drive masonry screws.

2 CLAMP THE CABLE TO THE SUBPANEL. Ask the building department what kind of feeder cable or conduit to use. Punch out the slug from a knockout. Install a cable clamp in the hole. Strip plenty of sheathing so that the wires can run around the perimeter of the subpanel. Clamp the cable to the box.

MAJOR PROJECTS

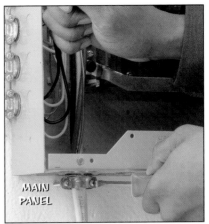

3 **CONNECT THE WIRES IN THE SUBPANEL.** Carefully bend the feeder wires so they run around the perimeter of the subpanel. Run the white wire to the main terminal on the neutral bar and run the ground wire to the ground bar. Run the red and black wires to each of the main terminals on the hot bars. Snip each wire to length and strip off ½-inch of insulation. Poke the wires into the terminals and tighten the setscrews.

4 **MAKE ROOM FOR THE FEEDER BREAKER IN THE MAIN PANEL.** Shut off the main breaker in the main service panel. If you do not have two open slots for the feeder breaker, replace four full-size breakers with two tandem breakers. You may have to remove two breakers and the wires and later connect them to the subpanel.

5 **CLAMP THE CABLE TO THE MAIN PANEL.** Run the feeder cable to the service panel. Strip plenty of sheathing, punch out a knockout slug, and clamp the cable.

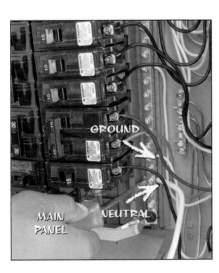

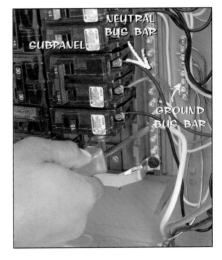

6 **CONNECT THE GROUND AND NEUTRAL WIRES AT THE MAIN PANEL.** Bend the ground wire and the neutral wire around the perimeter of the service panel to open terminals on the neutral bar, and snip them to length. Strip ½-inch of insulation from the neutral wire. Poke the wires into the terminals and tighten the setscrews.

7 **CONNECT THE WIRES TO THE FEEDER BREAKER AT THE MAIN PANEL.** Cut the red and black wires to length. Strip ½-inch of insulation from each and connect them to the two setscrew terminals of a double-pole feeder breaker. Snap the feeder breaker into place. Turn off the feeder breaker and turn on the main breaker.

8 **WIRE NEW BREAKERS TO THE SUBPANEL.** Run cable for new circuits into the subpanel. Connect the wires to new circuit breakers as you would in a main service panel (pages 168–169). Turn on the feeder breaker in the main panel to energize the subpanel.

MAJOR PROJECTS

Although homeowners sometimes install new service panels themselves, many building departments insist that the job be done only by licensed and bonded electricians. Because the job is complex and potentially dangerous, tackle it only if you have an excellent source of professional advice.

WHEN TO INSTALL A NEW PANEL

If your service panel is an old-fashioned fuse box, you may want to update it by installing a breaker panel. However, if your circuits rarely blow fuses, the box is not damaged, and you do not plan to add new circuits, there is no compelling reason to upgrade it.

If you need to add new circuits and the service panel cannot accept additional breakers, the easiest solution is to add a subpanel (pages 170–171). A professional electrician—who is used to working with live electricity and sorting out tangles of wires—may prefer to replace the old service panel with a larger one.

If your existing electrical service is insufficient—for example, if you have 60-amp service and need 100 amps or if you have 100-amp service and need 200 amps—you need a new service panel. You also may need to have the utility company change the wires that enter your home.

WHAT'S INVOLVED

An electrician or the utility company must first disconnect the power coming to the house—often by cutting live wires near the service entrance. The electrician can then provide temporary electrical service by tapping the live wires. Obviously, all this is too dangerous for a homeowner.

If your service amperage needs to be increased, the utility company may need to install thicker wires. If you have very old electrical service with only two wires, the utility company must run three wires to your home.

Replacing the service panel is now primarily a matter of managing a tangle of wires. All the wires running to the panel must be disconnected, tagged, and pulled out of the panel. Then the panel must be removed and another one mounted. It's important to position the new panel so that all the wires can reach the breakers and bus bars. When the wires are attached, power can be reconnected.

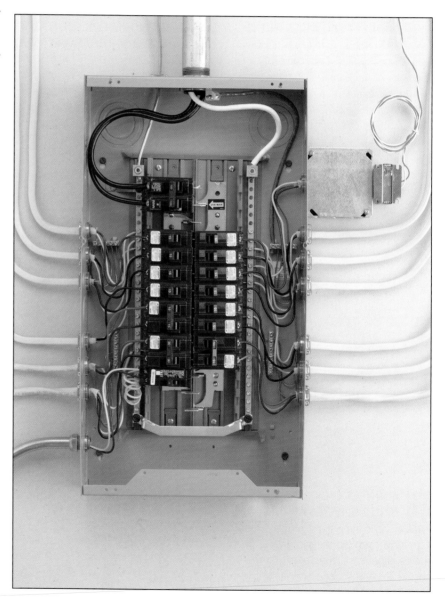

IT'S A TANGLE. Sorting out the incoming circuits and hooking them up properly is a job best left to the pros. It's not a highly technical job, but it does require clear thinking and the skills that come with practice. Hire a professional electrician to do this.

WIRING A BATHROOM

Bathrooms are usually small, with only a few electrical fixtures and devices. Because they are damp places, specific code requirements apply. You'll need at least two circuits—one for the lights and one for the receptacles.

A bathroom must have at least one ground-fault circuit interrupter (GFCI) receptacle on a 20-amp circuit. The receptacle must be within 12 inches of a sink. If the sink has two bowls, place a single receptacle between the bowls or put one receptacle on each side of the sink. Some codes allow bathroom receptacles to share a circuit with another receptacle elsewhere in the house.

Codes usually require a vent fan. Usually, a vent fan supplies light as well as ventilation. Unless the fan is very powerful or has a heating unit, a vent fan can share a circuit with other bathroom lights.

All overhead lights must be approved for moist rooms. Install lights over the sink, in the main area, and over the tub/shower. Codes may require that you install GFCI protection for the light circuit, especially to protect the light over the tub/shower. All bathroom receptacles must be GFCI protected.

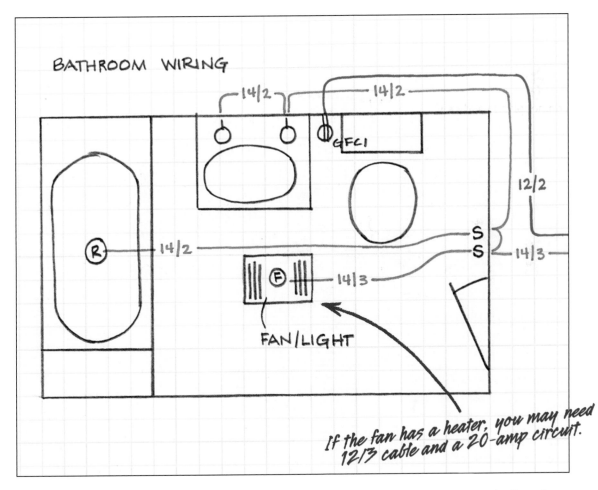

BATHROOM WIRING

14/2 14/2 GFCI 12/2 S S 14/3 14/2 R 14/3 F FAN/LIGHT

If the fan has a heater, you may need 12/3 cable and a 20-amp circuit.

A TYPICAL BATHROOM. Only one receptacle is usually needed in a bathroom. Here, a single GFCI is on its own 15-amp circuit. One 15-amp circuit supplies waterproof can lights over the tub/shower, the lights beside the mirror, and the fan/light. If you install a fan/light with a heating unit, it may pull as much as 1500 watts and will require a separate 20-amp circuit. Switches are conveniently positioned beside the door.

WIRING A KITCHEN

A kitchen is the room that has the most electrical devices and fixtures. It's not unusual to have eight or more circuits in a large kitchen, an organizational challenge.

- **Receptacles.** Position small-appliance receptacles over the counter no more than 4 feet apart and a couple of inches above the countertop backsplash. Have at least two 20-amp circuits for these. Plan the placement of the toaster, mixer, and other appliances. Avoid a potential tangle of cords. Islands and peninsulas also need appliance receptacles, which can be mounted on the sides of cabinets.

Older microwaves are heavy users of electricity. Many kitchens have a dedicated 20-amp circuit supplying the microwave receptacle. Most new microwaves use far less power and can be safely plugged into any dedicated small-appliance receptacle.

A refrigerator receptacle needs its own 15-amp circuit. Wire a split and switched receptacle (pages 143–144) for the garbage disposal, and place the switch on the wall above the countertop or on a base cabinet. This receptacle has an always-hot outlet that can be used for another appliance.

- **Appliances.** A dishwasher can be hard-wired, meaning you run cable directly into it. Hard-wire a range hood as well (pages 159–160). An electric range needs a 240/120-volt receptacle; however, a gas range needs only a 120-volt receptacle.
- **Lights.** A kitchen with many lights might need more than one 15-amp circuit; add up the kitchen's total wattage to find out. Position switches for maximum convenience. A large kitchen may need three-way switches.

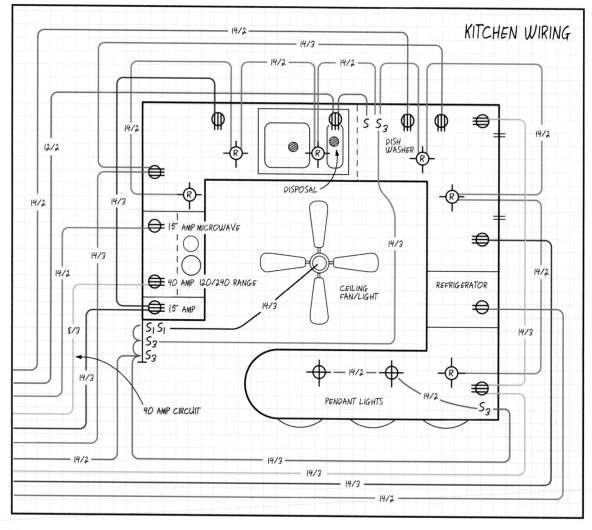

KITCHEN WIRING

A MODEST-SIZE KITCHEN. A 15-amp lighting circuit supplies a single ceiling fan/light, pendant lights, and recessed can lights; many lights are controlled by 3-way switches. The dishwasher and disposer share a circuit; the microwave and refrigerator each have their own circuit. Over-the-countertop receptacles are GFCI protected. If they're split (page 144), as required in some areas, none could be GFCI. The electric range has its own 50-amp, 240-volt circuit. (For a larger kitchen, see pages 114–115.)

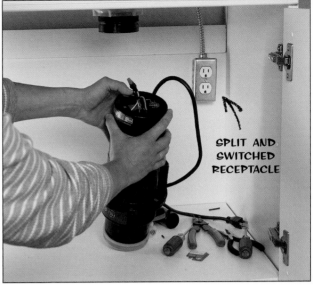

WIRING A DISHWASHER. If the dishwasher does not come with a plug, run two-wire cable into the space, leaving plenty of slack. Slide in the dishwasher and connect the plumbing. Remove the electrical cover and clamp the cable to the dishwasher electrical box. Splice white to white and black to black wires. Fold back the wires and snap on the cover.

WIRING A GARBAGE DISPOSAL. Install a receptacle box in the wall under the sink and a switch box in an easy-to-reach place above the countertop. Wire for a split and switched receptacle (pages 143–144). Remove the electrical cover from the disposal, strip the ends of an appliance cord, and wire the cord to the disposal. After completing plumbing connections, plug the disposal into the switched outlet.

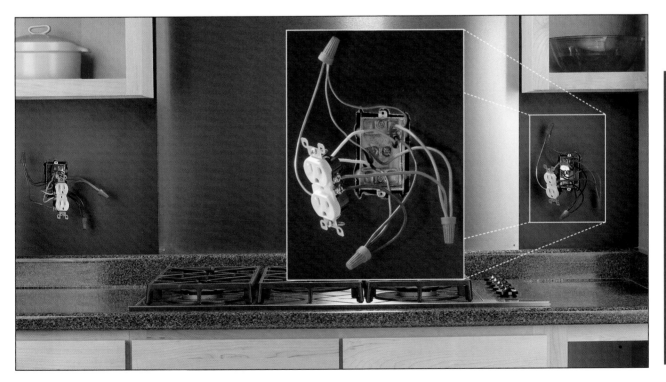

ADDING SMALL-APPLIANCE RECEPTACLES. Run 14/3 cable from a double-pole 15 amp breaker into the first receptacle box then on to the other boxes. To distribute electrical load, the split receptacle must be wired alternately (1st and 3rd, 2nd and 4th, etc.) depending, as always, on local code. Shown above is a split receptacle installation.

WIRING A BEDROOM

ost bedrooms have either an overhead switched light or one switched receptacle, plus a receptacle or two on each wall. You can go beyond the basic necessities and supply your bedroom with electrical service to outfit a small office or to add a few creature comforts. NOTE: AFCIs (page 168) will be required beginning January, 2002. For your safety, install one for each bedroom circuit.

- **Receptacles.** Codes typically allow bedroom receptacles to be up to 12 feet apart. If you cut this distance in half, you'll improve receptacle accessibility and give yourself more options for arranging bedroom furniture. To provide a computer with maximum protection against power surges, wire an isolated-ground receptacle. For comfortable TV viewing while in bed, install a wall bracket for a TV with a nearby

receptacle, about 6½ feet above the floor.

Avoid placing a receptacle directly below a window: it may get wet if the window is open during a rainstorm. If you use a window air-conditioner, install a receptacle near the window. An average window unit does not pull heavy amperage, so you can use a 15-amp receptacle on the same circuit as the rest of the bedroom receptacles. A heavy-duty air-conditioner may need a dedicated 20-amp receptacle.

- **Lights.** To control an overhead fan/light, run three-wire cable from the ceiling box to the switch box and install a fan/light switch (page 36). Consider installing three-way switches at the door and by the bed for convenience. Or, install a remote-control switch (page 98). Place a reading lamp at both sides of the bed, each with its own switch.

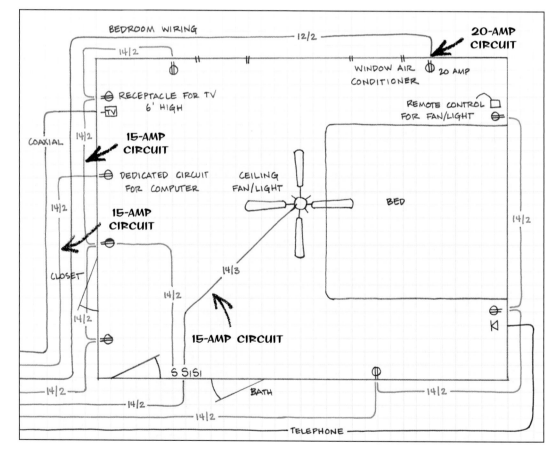

A MULTI-USE BEDROOM. Separate reading lights on each side of the bed each have conveniently placed switches. A receptacle with its own dedicated circuit guards a computer against damage caused by power surges (page 101). A receptacle 6 feet from the floor supplies power to a wall-mounted TV and VCR, and eliminates unsightly dangling cords. The fan/light is controlled by a wall switch and a remote control. (Wire the fan and light separately, using the three-way wiring described on pages 148–150.) A receptacle placed higher than usual near the window accommodates a window air-conditioner.

WIRING A LAUNDRY ROOM

In a laundry room, receptacles that feed the washing machine, gas dryer, and other appliances must be on 20-amp circuits that are not used by any other room. The receptacles must be ground-fault circuit interrupter (GFCI) protected.

If the dryer is electric, you also will need a 30-amp, 120/240-volt receptacle. Use 10-gauge wire and connect the dryer directly to a 30-amp, 240-volt breaker or fuse (page 145). In this example, the washer is on its own circuit. A washing machine could share a circuit with a gas dryer.

Because these machines vibrate, fasten the wiring securely. Local codes may allow NM or armored cable, but conduit is more secure. See pages 126–127 for conduit installation instructions.

Laundry room lights don't need their own circuits. But you shouldn't put them on circuits other than receptacle circuits so you won't be without light if a faulty appliance causes a circuit overload.

CLOSER LOOK

LIGHTING CLOSETS AND STORAGE SPACES

The days of exposed lightbulbs on pull-chain switches are past. Lights in closets, attics, crawlspaces, and other storage areas must now be recessed or enclosed, controlled by wall switches, and positioned at least 18 inches away from flammable materials. Wherever there is equipment that must be serviced—such as a sump pump or a water heater—there must be a light controlled by a wall switch.

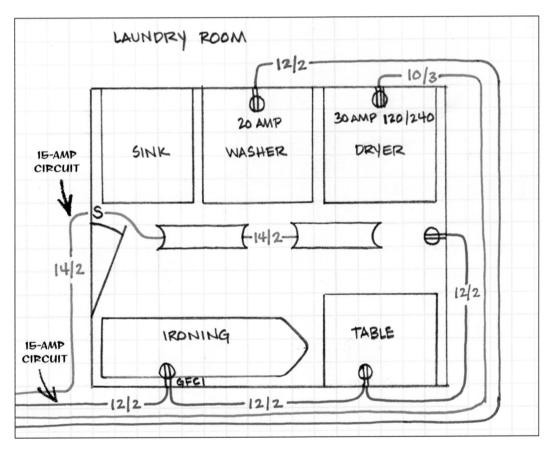

A HARD-WORKING LAUNDRY AREA. The dryer has a dedicated 30-amp, 240-volt circuit, and the washer has its own 15-amp, 120-volt circuit. Because laundry room wiring may be specified by local codes, check with your local building authority before proceeding. Fluorescent lights are on a 15-amp circuit, which they may share with lights in other rooms.

LOW-VOLTAGE WIRING

In addition to having miles of standard-voltage wiring, your home likely has hundreds of yards of thin wires that carry little or no power. These wires lead to telephones, thermostats, door chimes, VCRs, and TVs.

These wires and cable carry voltage that is so low it cannot harm you. Still, you should respect low-voltage wiring. Once the wiring is damaged, it can be difficult to diagnose the problem and to repair it. Hide low-voltage wiring inside walls or behind moldings when possible. If wires must be exposed, pull them taut and staple them firmly.

CHAPTER TWELVE PROJECTS

LOW-VOLTAGE WIRING

Homer's Hindsight

DRILL NEW HOLES

When running a new telephone line through my basement, I saved a little time by threading the cable through the same hole as was used by electrical cable. The result: static and buzzing on the line. Always keep phone and TV cable 2 inches apart from electrical cable when running parallel, and 1 inch apart wherever they cross.

INSTALLING TELEPHONE WIRING

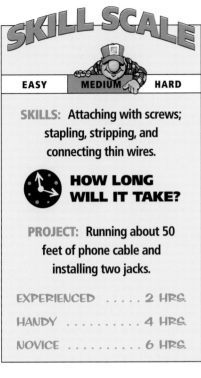

SKILL SCALE

EASY | MEDIUM | HARD

SKILLS: Attaching with screws; stapling, stripping, and connecting thin wires.

HOW LONG WILL IT TAKE?

PROJECT: Running about 50 feet of phone cable and installing two jacks.

EXPERIENCED . . . . . 2 HRS.

HANDY . . . . 4 HRS.

NOVICE . . . . 6 HRS.

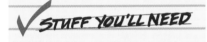

STUFF YOU'LL NEED

TOOLS: Drill, screwdriver, lineman's pliers, combination strippers

MATERIALS: Solid-core telephone cable, phone jacks, staples

WORK SMARTER

FRAGILE WIRES

Category 5 cable and telephone wire are fragile. Don't bend, flatten, or otherwise compromise these wires. A damaged wire can result in a distorted connection, especially for computers.

Adding a new telephone jack is straightforward work. Just run cable and connect wires to terminals labeled with their colors. The most difficult part is hiding the cable.

Depending on your service arrangement, it may be less expensive to have the phone company install new service for you. The lines they install will be under warranty—all future repairs will be free.

Cheap telephone cable has wires that are difficult to strip and splice. Spend a little more for "24-AWG" cable with solid-core wires.

Make all connections in a jack or junction box. Plan cable paths so as little of the cable as possible can be seen. For instance, going through a wall (page 180) saves you from running unsightly cable around door moldings. Use these same techniques to run speaker wire.

BUYER'S GUIDE

CATEGORY 5 CABLE

WIRE FOR THE FUTURE

As long as you're running phone lines, spend a little more for Category 5 cable, which can handle connections for Internet and high-speed data networks as well as standard telephone connections.

**1 OPTION A:
TAP INTO A PHONE JACK.**
Unscrew the cover from a phone jack or a phone junction box. Strip about 2 inches of sheathing and ½ inch insulation from each wire. (Standard phones use only two of the wires, but it doesn't hurt to connect all the wires.) Loosen each terminal screw. Bend the wire end in a clockwise loop, slip it under the screw head, and tighten the screw.

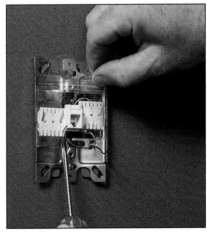

**OPTION B:
USE PUSH-ON CONNECTORS.** Some jacks have terminals that clamp onto the wire so you don't have to strip it. Just push the wire down into the slot until it snaps into place.

LOW-VOLTAGE WIRING

179

2 **HIDE CABLE.** Use any trick you can think of to tuck away unsightly cable. Pry moldings away from the wall, slip the cable in behind, and renail the molding. Or, pull carpeting back one short section at a time, run cable along the floor behind the tack strip, and push the carpet back into place.

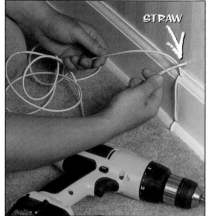

STRAW

3 **RUN CABLE THROUGH A WALL.** To go through a wall, drill a hole using a long, ¼-inch drill bit, then insert a large drinking straw. Fish the cable through the straw. When you're finished, split and remove the straw.

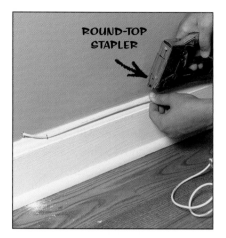

ROUND-TOP STAPLER

4 **STAPLE EXPOSED CABLE.** When there is no choice but to leave cable exposed, staple it in place every foot or so along the top of the baseboard. Use a round-top stapler or plastic-shielded staples that hammer into place. (Square-cornered staples damage the cable sheathing.)

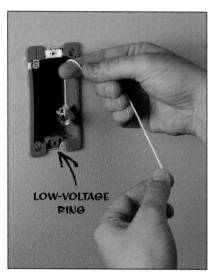

LOW-VOLTAGE RING

5 **INSTALL A WALL BOX.** A wall jack can attach to a low-voltage ring (as shown) or to an electrical remodel box. Cut a hole in the wall and install the ring. Tie a small weight to a string and lower it through the hole until you feel it hit the floor.

6 **PULL THE CABLE.** Drill a ⅜-inch hole at the bottom of the wall where you want the wire to go. Bend a piece of wire into a hook, slip it into the hole, and pull out a loop of the string. Tape the string to the cable and pull the cable up through the remodel box.

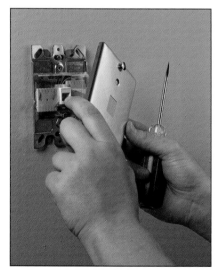

7 **INSTALL A WALL JACK.** Attach the base of the jack and make the connections. Install the cover plate.

LOW-VOLTAGE WIRING

180

RUNNING COAXIAL CABLE

Cable TV companies will run new lines and install jacks. Some do simple installations for free; for longer runs, they may charge and may not hide as much of the cable as you like. They also may increase your monthly fee after installing a second or third jack. Still, it's worth checking out the service options before doing your own installations.

Purchase RG6 coaxial cable for all runs through the house. Don't use RG59, which has less-substantial wire wrapping. Coaxial cable is thick and ugly, so fish it through walls when possible (pages 130–132).

If your cable signal is weak after adding new lines, install a signal booster to solve the problem. The booster attaches to the coaxial cable and plugs into a receptacle.

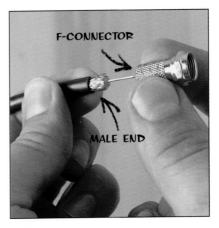

1 **MAKE A MALE END.** Use combination strippers to strip ¾ inch of insulation, exposing the bare wire. Do not bend the exposed wire. With a knife, carefully strip ⅜ inch of the thin outer sheathing only—do not cut through the metal mesh wrapping. Firmly twist a screw-on F-connector. For best results, purchase a special coaxial crimping tool and attach a crimp-on F-connector.

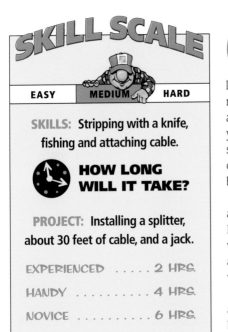

2 **SPLIT A LINE.** Cut the line you want to tap into. Install male ends on both ends of the cut line and the end of the new line. Insert and twist all three male connectors onto a signal splitter. Anchor the splitter with screws.

3 **INSTALL A JACK.** Cut a hole in the wall and run cable to it using the technique shown on page 180. (A regular electrical box can be used, though a low-voltage ring is preferable. See opposite page.) Strip the insulation to make a male end in the cable. Clamp the mounting brackets in the hole. Attach the cable end to the back of the jack by twisting on the F-connector. Tighten the connection with pliers or a wrench. Attach the jack to the wall by driving screws into the mounting brackets.

LOW-VOLTAGE WIRING

TROUBLESHOOTING A DOOR CHIME

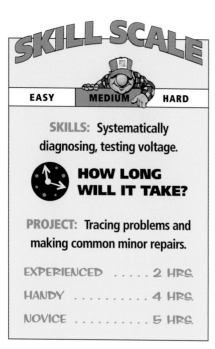

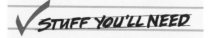

✓ STUFF YOU'LL NEED

TOOLS: Screwdriver, multitester, combination strippers

MATERIALS: The faulty component: button, chime, transformer, or bell wire

A doorbell or chime system is supplied with low-voltage power—between 8 and 24 volts—by a transformer. When the button is pressed, the circuit closes and sends power to the chime or bell.

FIXING COMMON PROBLEMS

Because the voltage associated with doorbells and chimes is low, there is no need to shut off power unless you are working on the transformer. Here's how to troubleshoot most problems.

■ **If a bell or chime develops a fuzzy sound,** remove the chime cover and vacuum out any dust and debris and brush off the bell or chimes.

■ **If you get only one tone** when the front (or only) button is pushed, check the wiring in the chime to see that the button is connected to the "front." On many two-button systems, the chime is supposed to "ding dong" when the front button is pushed, and only "ding" when the rear button is pushed.

■ **If the chime suddenly stops working** at the same time you blow a fuse or trip a breaker, restore power to the circuit supplying the transformer.

■ **If the chime stops working altogether,** conduct a systematic investigation, moving from the simplest to the most complex repairs. First check out the button(s), then the chime, and then the transformer. If none of these reveals a problem, the wiring may be damaged.

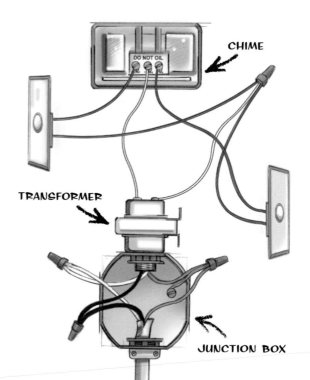

A TYPICAL TWO-BUTTON SETUP. The transformer—usually located in an out-of-the-way spot such as the basement, crawlspace, or cabinet interior—sends low-voltage power to the chime. There, one wire is connected to the chime. Another wire is spliced to two different wires, each of which travels through a button and back to the chime. When either button is pressed, the circuit is completed, power travels to the chime, and the chime rings.

1 **EXAMINE THE BUTTON.** Remove the screws while holding the button in place, and gently pull out the button. (Make sure the wires do not slide back into the hole.) Use a toothbrush to clean away any debris, cocoons, or corrosion, and tighten the screws. If either wire is broken, restrip and reconnect it. Retest the button.

TOUCH WIRES

2 **TOUCH WIRES TOGETHER.** If the button still doesn't work, loosen the terminal screws and remove the wires. Holding each wire by its insulation, touch the bare ends together. If the chime sounds, the button is faulty and needs to be replaced. If you see or hear a tiny spark but the chime does not sound, the chime may be faulty (Step 3). If there is no sound and no spark, check the transformer (Step 4, page 184).

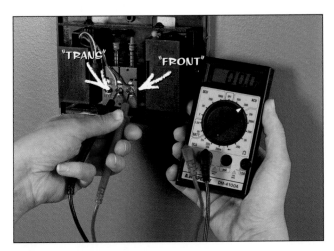

"TRANS"   "FRONT"

3 **TEST THE CHIME.** Remove the chime cover and ensure that all the wires are securely connected to terminals. Vacuum out any dust and scrape away any corrosion near the terminals. When you pull back a plunger and release it, the chime should sound. If not, clean any greasy buildup that may be gumming up the springs. If the chime still does not work, touch the probes of a multitester to the "front" and "trans" terminals, and to the "rear" and "trans" terminals. If power is present within two volts of the chime's printed voltage rating when you have someone press the chime's doorbell button, then the wiring is working fine and the chime itself is faulty and should be replaced.

BUYER'S $ GUIDE

REPLACING A CHIME
Purchase a chime with the same voltage rating as your transformer. It should be at least as large as the old chime so that you don't have to paint the wall around it. Label the wires with pieces of tape, unscrew the terminal screws, and remove them. Remove the screws holding the chime to the wall and pull it away. Thread the wires through the new chime and fasten the chime to the wall. Connect the wires to the terminals.

LOW-VOLTAGE WIRING

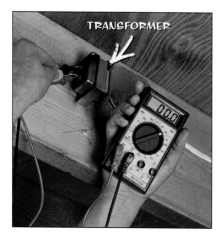

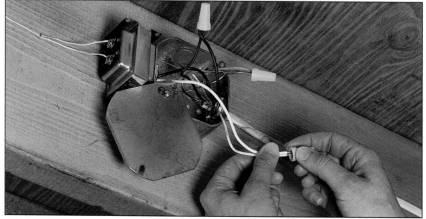

4 **TEST THE TRANSFORMER.**
Look for an exposed electrical box with the transformer attached. Tighten loose connections. Touch the probes of a multitester to both transformer terminals. If you get a reading of more than 2 volts below the transformer rating, the transformer is faulty and should be replaced.

5 **REPLACING A TRANSFORMER.**
Purchase a transformer with the same voltage rating as the old one. Shut off power to the circuit and open the adjacent junction box. Label the bell wires and disconnect them. Disconnect the transformer leads inside the junction box and disconnect the transformer. Thread the new transformer leads into the junction box, fasten the transformer to the box, and splice the leads to the wires. Connect the bell wires and test.

INSTALLING WIRELESS CHIMES

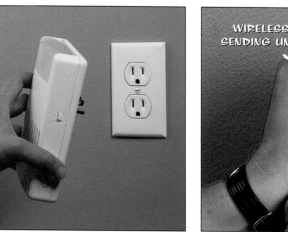

INSTALLING THE CHIME. Rather than going through the trouble of replacing defective bell wire, buy a wireless chime system. Installation is simple: Plug the chime into a standard receptacle., power the button with a battery, and attach the button to the house.

ADDING A WIRELESS CHIME TO AN EXISTING CHIME SYSTEM. If you can't hear your door chime everywhere in your home, add a wireless chime to your wired system. Remove the cover from the existing chime and loosen the terminal screws. Take the leads of the wireless chime's sending unit and insert them under the screws. Tighten the screws. Using its double-sided tape, stick the sending unit to the chime housing. Plug the wireless chime into a receptacle.

LOW-VOLTAGE WIRING

TROUBLESHOOTING A THERMOSTAT

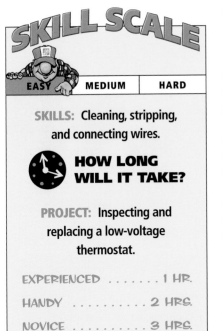
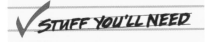

STUFF YOU'LL NEED

TOOLS: Small brush, screwdriver, combination strippers

MATERIALS: Short length of wire, replacement thermostat, scrap of bond paper

The round, low-voltage unit featured in most of these pictures is the most common type of thermostat in use. Yours may be rectangular, but its functions are the same.

If your furnace or air-conditioner fails to operate, check the thermostat for simple mechanical problems. The cover may be jammed in too far, disrupting the mechanism. A wire may have broken or come loose. Or, the parts may be covered with dust, inhibiting electrical contact.

If cleaning and adjusting do not solve the problem, replacing a thermostat is a simple job. Consider installing a programmable unit for more control options. Remember that a thermostat contains mercury, so dispose of it properly.

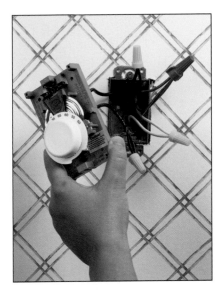

CHECKING A LINE-VOLTAGE THERMOSTAT. If your thermostat uses household current, always shut off power to the circuit before pulling it out. If it fails, disconnect it and take it to a dealer for service or replacement.

REVIEWING THE ANATOMY OF A LOW-VOLTAGE THERMOSTAT. Thin wires come from a transformer and connect to the thermostat base. You'll probably find one wire for the transformer, one for heat, one for air-conditioning, and one for a fan. (A heat pump uses six or more wires and has a special thermostat. Contact a dealer for repairs.) To protect circuitry, shut off power before you start to work.

WORK SMARTER

SEAL OFF DRAFTS

Even if your thermostat is on an interior wall, air coming through a hole behind it may throw its temperature readings out of whack, resulting in erratic heating. Remove the thermostat base from the wall and fill the hole with insulation or caulk.

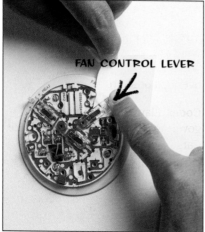

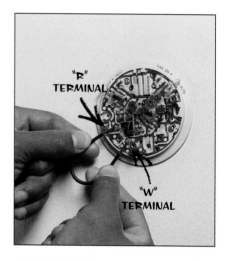

CLEANING THE CONTACTS WITH A BRUSH. Pull off the outer cover and use a soft, clean, dry brush to remove dust from the bimetal coil. Turn the dial to get at all the nooks and crannies.

CLEANING THE SWITCH CONTACTS. Remove the screws holding the thermostat body and pull out the body. Gently pull back on the fan control lever, slip a piece of white bond paper behind it, and slide the paper back and forth to clean the contact behind it. Do the same for the mode control lever, if there is one.

CONDUCTING A HOT-WIRE TEST. If heat does not come on, test to see if power is getting to the thermostat. Cut a short length of wire and strip both ends. Holding only the insulated portion, touch the bare ends to the terminals marked W and R. If the heating system starts to run, replace the thermostat. If nothing happens, troubleshoot or replace the transformer (page 184).

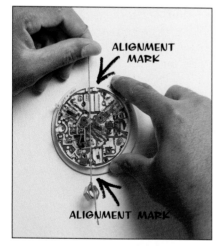

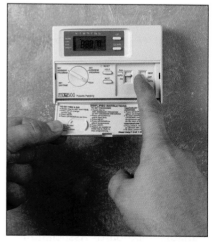

LEVELING A THERMOSTAT. If the temperature is always warmer or cooler than the thermostat setting, the thermostat may be out of level. Hold a level or a weighted string in front of the thermostat to see if the two alignment marks line up. If not, remove the mounting screws, realign the thermostat, and drive new screws.

REPLACING A LOW-VOLTAGE THERMOSTAT. Loosen the terminal screws and pull out the wires. Remove the mounting screws and pull out the plate. Clip the wires so they cannot slide back through the hole. Thread the wires through the new thermostat and hook the wires to the terminals. Check for leveling and attach the base to the wall with screws.

INVESTING IN A PROGRAMMABLE THERMOSTAT. Spend a little more and save money in the long run with a thermostat that adjusts heating or cooling several times a day.

LOW-VOLTAGE WIRING

GLOSSARY

Amp. A measurement of the amount of electrical current in a circuit at any moment. *See also* Volt *and* Watt.

Antioxidant. A paste applied to aluminum wires to inhibit corrosion and maintain safe connections.

Armored cable. Two or more insulated wires wrapped in a protective metal sheathing.

Ballast. A transformer that regulates the voltage in a fluorescent lamp.

Bell wire. A thin, typically 18-gauge wire used for doorbells.

Box. A metal or plastic enclosure within which electrical connections are made.

Bus bar. A main power terminal to which circuits are attached in a fuse or breaker box. One bus bar serves the circuit's hot side; the other, the neutral side. Some service panels and all subpanels have separate neutral and ground bus bars.

BX. Armored cable containing insulated wires but no ground wire.

Cable. Two or more insulated wires wrapped in metal or plastic sheathing.

Canadian Electrical Code (CEC). A set of rules governing safe wiring methods. Local codes sometimes differ from and take precedence over the CEC.

Canadian Standards Association (CSA). An independent testing agency that examines electrical components for safety hazards.

Canadian Underwriters Laboratories (CUL). An independent testing agency that examines electrical components for safety hazards.

Circuit. The path of electrical flow from a power source through an outlet and back to ground.

Circuit breaker. A switch that automatically interrupts electrical flow in a circuit in case of an overload or short.

Codes. Llaws and regulations governing safe wiring practices. *See* Canadian Electrical Code.

Common. A terminal on a three-way switch, usually with a dark-colored screw and marked COM.

Conductor. A wire or anything else that carries electricity.

Conduit. Rigid (metal or PVC) or flexible plastic tubing through which wires are run.

Continuity tester. An instrument that tells whether a device is capable of carrying electricity.

Dimmer. A rotary or sliding switch that lets you vary the intensity of a light.

Duplex receptacle. A device that includes two plug outlets. Most receptacles in homes are duplexes.

Electrical metallic tubing (EMT). Thin-walled, rigid conduit suitable for indoor use.

End-of-the-run. A device located at the end of a circuit. No wires continue from its box to other devices.

Feed wire. A wire that brings household current to a device.

Fishing. Pulling cables through finished walls and ceilings.

Fish tape. A hooked strip of spring steel used for fishing cables through walls and for pulling wires through conduit.

Fixture. Any light or other electrical device permanently attached to a home's wiring.

Flexible metal conduit. Tubing that can be easily bent by hand. *See also* Greenfield.

Fluorescent tube. A light source that uses an ionization process to produce ultraviolet radiation. This radiation becomes visible light when it hits the coated inner surface of the tube.

Four-way switch. A type of switch used to control a light from three or more locations.

Fuse. A safety device designed to stop electrical flow if a circuit shorts or is overloaded. Like a circuit breaker, a fuse protects against fire from overheated wiring.

Ganging. Assembling two or more electrical components into a single unit. Boxes, switches, and receptacles are often ganged.

Greenfield. Flexible metal conduit through which wires are pulled.

Ground. Refers to the fact that electricity always seeks the shortest possible path to the earth. Neutral wires carry electricity to ground in all circuits. An additional grounding wire, or the sheathing of metal-clad cable or conduit, protects against shock from a malfunctioning device.

Ground-fault circuit interrupter (GFCI). A safety device that senses any shock hazard and shuts off a circuit or receptacle.

High-intensity discharge (HID). A type of lighting, including lamps such as halogen, mercury vapor, metal halide, and sodium. All HIDs produce a bright, economical light.

Hot wire. The conductor of current to a receptacle or other outlet. *See also* Neutral wire *and* Ground.

Incandescent bulb. A light source with an electrically charged metal filament that burns at white heat.

GLOSSARY (CONTINUED)

Insulation. A nonconductive covering that protects wires and other electricity carriers.

Junction box. An enclosure used for splitting circuits into different branches. In a junction box, wires connect only to each other, never to a switch, receptacle, or fixture.

Kilowatt (kW). One thousand watts. A kilowatt hour is the standard measure of electrical consumption.

Knockouts. Tabs that can be removed to make openings in a box for cable or conduit connectors.

LB connector or fitting. An elbow for conduit with access for pulling wires. Connections cannot be made within this fitting.

Lead. A short wire coming from a fixture, typically stranded, to which a household wire is spliced. It is used instead of a terminal.

MC cable. Armored cable containing at least two insulated wires and an insulated ground wire.

Middle-of-the-run. A device located between two other devices on a circuit. Wires continue from its box to other devices.

Multitester. A device that measures voltage in a circuit and performs other tests.

Neutral wire. A conductor that carries current from an outlet back to ground, clad in white insulation. *See also* Hot wire *and* Ground.

New-work box. A metal or plastic box attached to framing members before the wall material is installed.

Nonmetallic (NM) sheathed cable. Two or more insulated wires and a bare ground wire clad in a plastic covering.

Outlet. Any potential point of use in a circuit, including receptacles, switches, and light fixtures.

Old-work box. *See* Remodel box.

Overload. A condition that exists when a circuit is carrying more amperage than it was designed to handle. Overloading causes wires to heat up, which in turn blows fuses or trips circuit breakers.

Pigtail. A length of wire, stripped at both ends, spliced with one or more other wires. It is used instead of attaching two or more wires to a terminal, an unsafe connection.

Polarized plugs. Plugs designed so the hot and neutral sides of a circuit can't be accidentally reversed. One prong of the plug is a different shape than the other.

Raceway wiring. Surface-mounted channels for extending circuits.

Receptacle. An outlet that supplies power for lamps and other plug-in devices.

Recessed can light. A light fixture set into a wall cavity so the lens and trim are flush with the ceiling.

Remodel box. A metal or plastic box, sometimes called an "old-work" box, designed for a hole cut in drywall or plaster and lath.

Rigid conduit. Wire-carrying metal tubing that can be bent only with a special tool.

Romex. A trade name for nonmetallic sheathed cable. *See* Nonmetallic sheathed cable.

Service entrance. The point where power enters a home.

Service panel. The main fuse box or breaker box in a home.

Short circuit. A condition that occurs when hot and neutral wires contact each other. Fuses and breakers protect against fire, which can result from a short.

Stripping. Removing insulation from wire, or sheathing from cable.

Subpanel. A subsidiary fuse box or breaker box linked to a service panel that has no room for additional circuits.

System ground. A wire connecting a service panel to the earth. It may be attached to a main water pipe, to a rod driven into the ground, or to a plate embedded along a footing.

Three-way switch. Operates a light from two locations.

Time-delay fuse. A fuse that does not break the circuit during the momentary overload that can happen when an electric motor starts up. If the overload continues, the fuse blows, shutting off the circuit.

Transformer. A device that reduces or increases voltage. In home wiring, transformers step down current for use with low-voltage equipment such as thermostats and doorbell systems.

Travelers. Two of the three conductors that run between switches in a 3-way installation.

Underwriters knot. A knot used as a strain relief for wires in a lamp socket.

Volt. A measure of electrical pressure. Volts × amps = watts.

Watt. A measure of the power an electrical device consumes. *See also* Volt, Amp, *and* Kilowatt.

Wire nut. A screw-on device used to splice two or more wires.

INDEX

A

AC cable, 16
accent lights
   described, 93
   living area, 91, 92
adapters, grounding, 53, 107
AFCI, 168, 176
aluminum wire, 64
amps
   appliance ratings, 63
   circuit capacities, 62
   defined, 8
   service panels, 60, 116–117
   subpanels, 170–171
anywhere switches, 98
appliances
   amp/watt ratings, 63
   cords, 73
   receptacles, 18, 76–77, 145, 174–175
   wiring, 174–175
arc-fault circuit interrupters. *See* AFCI
armored cable
   anchoring, 120
   described, 16
   working with, 124–125
armored cable cutters, 111, 124
attics
   fans, 153, 155
   lights, 177
   running cable, 130, 132

B

ballasts, 78–79, 104
basements, running cable, 130, 163
bathrooms
   code, 113
   lighting plan, 90
   vent fans, 156–158
   wiring, 173
bedrooms
   code, 113
   lighting plan, 92
   wiring, 176
boxes. *See also* junction boxes
   ceiling fan, 32–36
   code, 113
   GFCI, 27
   junction, 58, 137
   plastic vs. metal, 118–119
   recessed, 82
   remodeling, 133–134
   repairing wires in, 82
   safety inspections, 54, 58–59

sizing, 116
   in unfinished framing, 128
breaker boxes, 14–15, 117. *See also*
   service panels
breakers
   adding new, 168–169
   AFCI, 168, 176
   GFCI, 100
   resetting, 80
   types, 117, 168
bus bars, 14–15
BX cable, 36

C

cabinets, lights in, 92
cable. *See also* specific types
   anchoring, 120
   code, 113
   in finished walls, 130–132
   for outdoor lines, 164–165
   safety inspections, 57
   telephone, 179–180
   types, 16
   for under-cabinet lights, 30
   in unfinished framing, 128–129
cable clamps, 123
cable rippers, 122
cable TV, installing, 181
can lights. *See* recessed lights
category 5 cable, 16, 179
ceiling fans
   boxes, 32–34
   dimmers, 25
   hanging, 32–36
   remote controls, 35
   safety, 33
   switches, 34, 35
ceiling lights
   adding wall switch to, 142
   fixture types, 84–85
   raceway installation, 139
   safety, 33
   with switches, adding, 146–147
   upgrading, 28–29
chandeliers, 84
channel-joint pliers, 111
circuit breakers, 14, 80
circuit finders, 43–45
circuits
   adding new, 168–169
   code, 113
   described, 8–10
   loads, calculating, 62–63, 116–117
   mapping, 61, 114–115

   for outdoor lines, 164–165
   overloads, avoiding, 62–63
   overloads, results of, 15
   in power distribution system, 8–10
   shutting off power, 15
   types, 10
closet lights, 55, 105, 177
coaxial cable
   described, 16
   installing, 181
coaxial crimpers, 111
coaxial strippers, 111
codes
   common requirements, 112–113
   fluorescent fixtures, 104
   GFCI receptacles, 27, 113
   grounding, 12–13
   meeting, 52
   outdoor receptacles, 162
   stairways, 148
   240-volt circuits, 10
   240-volt receptacles, 18
   under-cabinet lights, 30–31
   when to consult, 52
color coding
   wire, 16, 40
   wire nuts, 50
combination strippers, 42, 46
conduit
   anchoring, 120
   described, 16
   outdoor, 164–165
   running, 126–127
conduit reamers, 111
connections, 9, 48–49
continuity testers, 42–45
contractors, electrical, 52
cord channel, 68–69
cords
   appliance, 73
   lamp, 68–70
   safety inspections, 56
cord switches, 73
cove lighting, 31, 89
crawlspaces, lights in, 177

D

deck lights, 96
designer tips
   cove lighting, 31
   recessed lights, 38
   rope lights, 105
desk lamps, 70
diagonal cutters, 42

QUICK GUIDE TO CABLE SELECTION
NOTE: INFORMATION BELOW APPLIES TO COPPER WIRE ONLY

▽ = Indoor NM-B with Ground

▽ = Outdoor UF-B with Ground

▼ = Amperage

Confirm the right cable for your project using the illustration above. Distance may impact the appropriate gauge of wire in the cable. Cables are labeled with two numbers. The first indicates wire size and the second how many wires are in the cable. For example, 12/2 cable means 12-gauge wire and two wires plus a ground wire.

CHOOSE THE RIGHT GAUGE WIRE

GAUGE GUIDE FOR MAXIMUM ALLOWABLE WIRE LENGTH, 2% NOMINAL VOLTAGE DROP

		COPPER CONDUCTORS							
AMPS	WATTS	UP TO 50'	UP TO 100'	UP TO 150'	UP TO 200'	UP TO 250'	UP TO 300'	UP TO 400'	UP TO 500'
15	1800	14 Gauge	12 Gauge	10 Gauge	6 Gauge	6 Gauge	6 Gauge	4 Gauge	4 Gauge
20	2400	12 Gauge	10 Gauge	8 Gauge	6 Gauge	6 Gauge	4 Gauge	4 Gauge	2 Gauge
30	3000	10 Gauge	8 Gauge	6 Gauge	6 Gauge	4 Gauge	4 Gauge	2 Gauge	2 Gauge
40	4800	8 Gauge	6 Gauge	4 Gauge	4 Gauge	2 Gauge	2 Gauge	1 Gauge	1/0 Gauge
50	6000	6 Gauge	6 Gauge	4 Gauge	4 Gauge	2 Gauge	1 Gauge	1/0 Gauge	2/0 Gauge

This information is provided as a general guideline only. For safe wiring practices consult the National Electric Code®, local building codes and regulations, and your local building inspector. Always remember that installation of electrical wire can be hazardous and, if done improperly, can result in personal injury or property damage.

The correct wire gauge (size) for your project is based on the amperage, or power, required and the distance the wire travels. The chart at left helps you determine what gauge wire you need. Remember, the smaller the number, the larger the wire. For example, a 6-gauge wire is larger than an 8-gauge wire.

Chart and illustration provided and copyrighted by General Cable Technologies Corporation, makers of Romex® brand cable.

Canadian codes differ from the above charts, and typically require a 14/2 cable on most household circuits. Check your provincial codes before completing your project.